AF371868

Representation Theory of Finite Groups: Algebra and Arithmetic

Steven H. Weintraub

Graduate Studies
in Mathematics

Volume 59

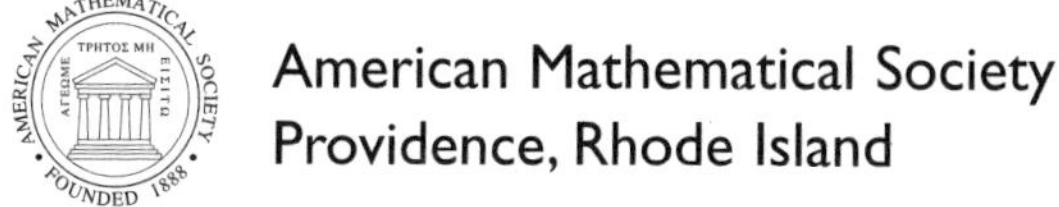

American Mathematical Society
Providence, Rhode Island

Editorial Board

Walter Craig
Nikolai Ivanov
Steven G. Krantz
David Saltman (Chair)

2000 *Mathematics Subject Classification.* Primary 20C05, 20C15, 20C20;
Secondary 20–01.

For additional information and updates on this book, visit
www.ams.org/bookpages/gsm-59

Library of Congress Cataloging-in-Publication Data
Weintraub, Steven H.
   Representation theory of finite groups / Steven H. Weintraub
      p. cm. — (Graduate studies in mathematics, ISSN 1065-7339 ; v. 59)
   Includes bibliographical references and index.
   ISBN 0-8218-3222-0 (alk. paper)
   1. Representations of groups.   2. Finite groups.   I. Title.   II. Series.

QA176.W45   2003
512′.2—dc21
                                                                    2003045186

Copying and reprinting. Individual readers of this publication, and nonprofit libraries acting for them, are permitted to make fair use of the material, such as to copy a chapter for use in teaching or research. Permission is granted to quote brief passages from this publication in reviews, provided the customary acknowledgment of the source is given.

Republication, systematic copying, or multiple reproduction of any material in this publication is permitted only under license from the American Mathematical Society. Requests for such permission should be addressed to the Acquisitions Department, American Mathematical Society, 201 Charles Street, Providence, Rhode Island 02904-2294, USA. Requests can also be made by e-mail to `reprint-permission@ams.org`.

© 2003 by the American Mathematical Society. All rights reserved.
The American Mathematical Society retains all rights
except those granted to the United States Government.
Printed in the United States of America.

♾ The paper used in this book is acid-free and falls within the guidelines
established to ensure permanence and durability.
Visit the AMS home page at `http://www.ams.org/`

10 9 8 7 6 5 4 3 2 1   08 07 06 05 04 03

To Aunt Ros

Contents

Preface

This book is an introduction to group representation theory. With the exception of this preface, it presupposes no prior knowledge of the subject, not even what a representation is. The reader with no prior knowledge should have a look at Chapter 1, which presents an introduction to the theory.

In this preface we discuss our approach and objectives, so here the reader must have some prior knowledge to understand what is going on.

This is a text on the representation theory of finite groups. The theory can be divided into two parts: ordinary (or semisimple) representation theory, the case where the characteristic of the field is 0 or prime to the order of the group, and modular (or nonsemisimple) representation theory, the case where the characteristic of the field divides the order of the group.

In each case, our approach is to present the basic ideas of the theory in a reasonably general context. Thus we do not prove the basic results purely for group representations, but rather for the appropriate sort of rings and modules, deriving the results for group representations from them. This is the "Algebra" part of the subtitle: we do some general algebra motivated by its applications to representation theory. Also, particularly in our treatment of the characteristic 0 case, we do not just treat the case of algebraically closed fields, but rather pay quite a bit of attention to the question of the field of definition of a representation. This is the "Arithmetic" part of the subtitle.

Let us now be more specific. We begin, in Chapter 1, with an introduction to the representation theory of finite groups. Chapter 2 is a treatment of the theory of semisimple rings and modules. We begin Chapter 3 with some examples, but then derive the basic theorems of ordinary representation theory as immediate consequences of the results of Chapter 2. Of course, this theory is not just a special case of the more general theory, so we then develop the additional methods we need in this case, especially characters, the main (and powerful) computational tool here.

One of the most powerful methods in representation theory is that of induction, to which we devote Chapter 4. In the basics of induction, we are careful to adopt an approach that will generalize to the modular case. We conclude this chapter by proving Brauer's famous, and deep, theorem that all of the irreducible complex representations of a finite group of exponent u are defined over the cyclotomic field obtained by adjoining the u-th roots of unity to the field of rational numbers.

Next we turn to the modular theory, with a similar approach. We begin in Chapter 5 with a number of examples. Then in Chapter 6 we treat general rings and modules, and begin Chapter 7 by deriving the basic theorems of modular representation theory as immediate consequences of the results of Chapter 6. We then further develop modular representation theory and fully analyze the situation for some small groups.

The appendix contains several background results that we use and with which the reader may not be familiar.

The scope of this book can perhaps best be described by the following analogy. We explore widely in the valley of ordinary representations, and we take the reader over the mountain pass leading to the valley of modular representations, to a point from which (s)he can survey this valley, but we do not attempt to widely explore it. We hope the reader will be sufficiently fascinated by the scenery to further explore both valleys on his/her own.

This is a text, and we do not claim to have any new results here. Many of the proofs are standard, but part of the fun of writing this book has been the opportunity to think deeply about this beautiful subject, so many of the proofs are the author's own. This does not mean they are necessarily new, as the author may simply have rediscovered known proofs for himself, but, on the other hand, it does not mean that they are all necessarily already known, either.

The author would like to thank Springer-Verlag and Bill Adkins for permission to use material from [AW]: W. A. Adkins and S. H. Weintraub,

Algebra: An Approach via Module Theory, Graduate Texts in Mathematics volume 136, ©Springer-Verlag 1992 (corrected second printing 1999). Indeed, the author first started thinking seriously about group representation theory while writing [AW], which has a chapter on it, and has continued thinking about this subject over the past decade. Thus, Section 7.1 and Chapter 8 of [AW] are subsumed here, in some places taken over unchanged, but in most places reworked and deepened.

We have deliberately kept the prerequisites for this book to a minimum. The reader will require a sound knowledge of linear algebra and a good familiarity with basic module theory. As a text on module theory we naturally recommend [AW]. In addition, the reader will need to be familiar with some basic homological algebra: exact sequences, Hom and tensor product ([AW, sections 3.3 and 7.2]), and the definition and properties of a projective module ([AW, section 3.5]). The reader will also need to be familiar with the basics of field theory and algebraic number theory.

Some words about terminology and notation are in order. Two modules (or representations) are called distinct if they are not isomorphic. $A \subset B$ means that A is a proper subset of B and $A \subseteq B$ means that A is a subset of B. Boldface capital letters are reserved for fields, with $\mathbf{Q}$, $\mathbf{R}$, and $\mathbf{C}$ denoting the fields of rational, real, and complex numbers, respectively. The integers are denoted by $\mathbb{Z}$, and the cyclic group of order n by $\mathbb{Z}/n\mathbb{Z}$. All vectors are column vectors, but are not written that way for typographical reasons. Instead, $\mathbf{F}^n = \{[a_1, \dots, a_n] \mid a_i \in \mathbf{F}\}$ (note the use of square brackets) denotes the space of column vectors with entries in $\mathbf{F}$. The symbol $\square$ denotes the end of a proof; a proof consisting entirely of that symbol means the result follows easily (often immediately) from a previous result. Finally, we have adopted a conventional numbering system: Theorem a.b.c refers to Theorem b.c of Chapter a (or of the appendix, if a is A), and Theorem b.c refers to Theorem b.c of the current chapter.

Chapter 1

Introduction

We shall begin by giving some basic examples of the objects and questions we will be studying.

Definition 1.1. Let G be a group and $\mathbf{F}$ a field. An **F-representation** of G is a vector space V over $\mathbf{F}$ and a homomorphism $\sigma : G \to \mathrm{Aut}(V)$.

Concretely, to each element g of G we associate an invertible linear transformation $\sigma(g) : V \to V$, and these satisfy $\sigma(g)\sigma(h) = \sigma(gh)$ for any two elements g and h of G. If V is finite dimensional, we may choose a basis of V, and in that basis each $\sigma(g)$ is represented by a matrix $T(g)$, and these matrices satisfy $T(g)T(h) = T(gh)$. (We adopt the convention throughout that matrices always act on the left, on column vectors.)

Definition 1.2. Let $\sigma_i : G \to \mathrm{Aut}(V_i)$ be **F-representations** of G, $i = 1, 2$. Then σ_1 and σ_2 are **equivalent** if there is an isomorphism $\alpha : V_1 \to V_2$ such that $\alpha(\sigma_1(g))(v_1) = \sigma_2(g)(\alpha(v_1))$ for every $g \in G$ and every $v_1 \in V_1$.

In other words, σ_1 and σ_2 are equivalent if there is an isomorphism $\alpha : V_1 \to V_2$ making the diagram

$$
\begin{array}{ccc}
G \times V_1 & \xrightarrow{\ \sigma_1\ } & V_1 \\
{\scriptstyle id \times \alpha} \downarrow & & \downarrow {\scriptstyle \alpha} \\
G \times V_2 & \xrightarrow{\ \sigma_2\ } & V_2
\end{array}
$$

commute. Translated into matrix language, this means that there is an invertible matrix M such that $MT_1(g) = T_2(g)M$, i.e., $T_2(g) = MT_1(g)M^{-1}$, for each $g \in G$ (where $T_i(g)$ is the matrix of $\sigma_i(g)$ in the respective basis).

We will often refer simply to a representation of G (when $\mathbf{F}$ is understood) and call it σ (when V is understood) or V (when σ is understood).

Definition 1.3. Let V be a representation of G. The **degree** $\deg(V)$ of V is $\dim_{\mathbf{F}}(V) \in \{0, 1, 2, \dots\} \cup \{\infty\}$.

Definition 1.4. Let V be a representation of G. A subspace W of V is a **subrepresentation** of G if the restriction $\sigma \mid W : W \to W$ is a representation of G.

In other words, W is a subrepresentation of G if $\sigma(g) : W \to W$ for every $g \in G$.

Definition 1.5. Let $V \neq \{0\}$ be a representation of G. Then V is **irreducible** if it has no subrepresentation W other than $W = \{0\}$ or $W = V$.

Definition 1.6. Let $V \neq \{0\}$ be a representation of G. Then V is the **direct sum of representations** $V = W_1 \oplus W_2$ if $V = W_1 \oplus W_2$ as vector spaces and $\sigma(g)(w_1 + w_2) = \sigma_1(g)(w_1) + \sigma_2(g)(w_2)$ for every $w_i \in W_i$ and every $g \in G$ (where $\sigma : G \to \mathrm{Aut}(V)$ and $\sigma_i : G \to \mathrm{Aut}(W_i)$, $i = 1, 2$).

Definition 1.7. Let $V \neq \{0\}$ be a representation of G. Then V is **indecomposable** if $V = W_1 \oplus W_2$ implies $W_1 = \{0\}$ and $W_2 = V$ or vice versa.

Proposition 1.8. *If V is irreducible, then V is indecomposable.*

Proof. $\qquad\qquad\qquad\qquad\qquad\qquad\qquad\qquad\qquad\qquad\qquad\square$

Proposition 1.9.

(1) *If W is a subrepresentation of V, then V/W is a representation.*

(2) *If $V = W_1 \oplus W_2$, then V/W_1 is equivalent to W_2.*

Proof. $\qquad\qquad\qquad\qquad\qquad\qquad\qquad\qquad\qquad\qquad\qquad\square$

In matrix language, if V is reducible and W is a subrepresentation of V, then, choosing a basis for W and extending it to a basis for V, for every $g \in G$, $T(g)$ has the form

$$T(g) = \begin{bmatrix} T_1(g) & * \\ 0 & T_2(g) \end{bmatrix},$$

where $T_1(g)$ is the matrix of $\sigma_1(g)$, the representation of G on W, $*$ is an arbitrary matrix, and $T_2(g)$ is the matrix of $\sigma_2(g)$, the representation of G on V/W. If V is decomposable, $V = W_1 \oplus W_2$, then choosing a basis for V which is the union of bases for W_1 and W_2, for every $g \in G$, $T(g)$ has the form

$$T(g) = \begin{bmatrix} T_1(g) & 0 \\ 0 & T_2(g) \end{bmatrix}.$$

With this terminology in hand, we may formulate the basic question in the subject: *Find all $\mathbf{F}$-representations of G and classify them up to isomorphism.* We may break this question into pieces. We may first ask: *Find all* *indecomposable* **F**-*representations of* G (since every finite-dimensional representation of G is clearly a direct sum of indecomposables). This suggests the related question: *Find all irreducible* **F**-*representations of* G. We may then ask: *Find a method of determining whether two* **F**-*representations are equivalent.*

We may also ask about how representations change with the field. At this point it is easiest to formulate this question by using coordinates. If $\sigma : G \to \mathrm{Aut}(V)$ is an **F**-representation of V, then a choice of coordinates gives $\sigma : G \to \mathrm{Aut}(V) \to GL_n(\mathbf{F})$, the group of invertible n-by-n matrices with entries in **F**, where $n = \dim_{\mathbf{F}}(V)$. Then if **E** is an extension field of **F**, we have $GL_n(\mathbf{F}) \subseteq GL_n(\mathbf{E})$, so the composition $\sigma : G \to \mathrm{Aut}(V) \to GL_n(\mathbf{F}) \hookrightarrow GL_n(\mathbf{E})$ gives an **E**-representation $\widetilde{\sigma}$ of G. We ask: *If σ is irreducible or indecomposable, will $\widetilde{\sigma}$ be irreducible or indecomposable?* We also ask: *When does an* **E**-*representation ρ arise as $\widetilde{\sigma}$ for some* **F**-*representation σ?* In such a case we say that ρ is defined over **F**.

Finally, another question that we will investigate is: *What is the relationship between representations of a group and of its subgroups?* Clearly any representation σ of a group G gives a representation of any subgroup H of G (merely by restricting σ to H) but we would also like to see if it is possible to somehow build up representations of G from those of its subgroups.

Let us illustrate these general questions by looking at some particular examples. First note that for any group G and **F**-vector space $V \ne \{0\}$ we have a representation defined by $\sigma(g) = id : V \to V$ for all $g \in G$. In this case we say that G acts trivially on V or that σ is a **trivial representation**.

Notation. We remind the reader of our standing notational convention that $[a_1, \ldots , a_n]$ stands for the *column* vector with entries $a_1, \ldots , a_n$.

Example 1.10. Let $G = \mathbb{Z}/p\mathbb{Z}$, the cyclic group on p elements, p a prime. Choose a generator g of G, so $G = \{1, g, g^2, \ldots , g^{p-1}\}$.

(1) Let **F** be arbitrary and $V = \mathbf{F}^p = \{[a_1, \ldots , a_p]\}$. Define σ by

$$\sigma(g)([a_1, \ldots , a_p]) = [a_p, a_1, \ldots , a_{p-1}].$$

In the standard basis of $\mathbf{F}^p$, $\sigma(g)$ has matrix

$$\begin{bmatrix} 0 & 0 & & 0 & 1 \\ 1 & 0 & \cdots & 0 & 0 \\ \vdots & 1 & & \vdots & \vdots \\ \vdots & \vdots & & \vdots & \vdots \\ 0 & 0 & & 1 & 0 \end{bmatrix}.$$

(2) Let $W_1 = \{[a_1, \ldots, a_p] \in V \mid a_1 + \cdots + a_p = 0\}$. Note that W_1 has dimension $p - 1$ and is a subrepresentation of V. Thus V is not irreducible.

(3) Let $W_2 = \{[a_1, \ldots, a_p] \in V \mid a_1 = \cdots = a_p\}$. Note that W_2 has dimension 1 and is a trivial representation of G.

(4) Note that V/W_1 is isomorphic to W_2 and V/W_2 is isomorphic to W_1.

(5) If $\mathrm{char}(\mathbf{F}) \neq p$, then $V = W_1 \oplus W_2$. Thus in this case V is decomposable.

(6) If $\mathrm{char}(\mathbf{F}) = p$, then V is indecomposable, though it is not irreducible.

(7) Let $\mathbf{F}$ have characteristic p and let $U = \mathbf{F}^2 = \{[a_1, a_2]\}$. Give U the standard basis, and define σ by

$$[\sigma(g)] = \begin{bmatrix} 1 & 1 \\ 0 & 1 \end{bmatrix}.$$

Then U is indecomposable but not irreducible.

(8) Let $\mathbf{F} = \mathbf{C}$. For any integer t define $\theta_t : G \to \mathrm{Aut}(\mathbf{C})$ to be multiplication by $\exp(2\pi i/p)$. (Note this only depends on t modulo p.) Call the vector space on which θ_t acts X_t.

(9) Consider the representation W_1 of part (2) of this example as a representation over $\mathbf{Q}$. Then W_1 is irreducible.

(10) Consider the representation W_1 of part (2) of this example as a representation over $\mathbf{C}$. Then W_1 is equivalent to $X_1 \oplus \cdots \oplus X_{p-1}$. (In particular, W_1 is not irreducible.)

(11) As a representation over $\mathbf{C}$, the representation V of part (1) of this example is equivalent to $X_0 \oplus X_1 \oplus \cdots \oplus X_{p-1}$.

Example 1.11. Let $G = D_{2m}$, the dihedral group of order $2m$, $G = D_{2m} = \langle x, y \mid x^m = 1, y^2 = 1, xy = yx^{-1} \rangle$, and let $\mathbf{F} = \mathbf{C}$. Let $\zeta = \exp(2\pi i/m)$. Let $V = \mathbf{C}^2$ with the standard basis, and define φ_k by

$$[\varphi_k(x)] = \begin{bmatrix} \zeta^k & 0 \\ 0 & \zeta^{-k} \end{bmatrix} \quad \text{and} \quad [\varphi_k(y)] = \begin{bmatrix} 0 & 1 \\ 1 & 0 \end{bmatrix}.$$

Then φ_k and φ_ℓ are equivalent if and only if $k \equiv \pm\ell \pmod{m}$. Also, φ_k is irreducible unless $k = 0$ or, in case m is even, $k = m/2$.

For m odd, let $V = \mathbf{C}$ and define $\psi_+(x) = [1]$ and $\psi_+(y) = [1]$, and ψ_- by $\psi_-(x) = [1]$ and $\psi_-(y) = [-1]$. Then φ_0 is equivalent to $\psi_+ \oplus \psi_-$. For m even, let $V = \mathbf{C}$ and define ψ_{++} by $\psi_{++}(x) = [1]$ and $\psi_{++}(y) = [1]$, define ψ_{+-} by $\psi_{+-}(x) = [1]$ and $\psi_{+-}(y) = [-1]$, define ψ_{-+} by $\psi_{-+}(x) = [-1]$ and $\psi_{-+}(y) = [1]$, and define ψ_{--} by $\psi_{--}(x) = [-1]$ and $\psi_{--}(y) = [-1]$. Then φ_0 is equivalent to $\psi_{++} \oplus \psi_{+-}$, and $\varphi_{m/2}$ is equivalent to $\psi_{-+} \oplus \psi_{--}$.

The group D_{2m} is usually called a dihedral group only when $m \geq 3$. In case $m = 1$ or $m = 2$ the group D_{2m} is abelian, but the above discussion still goes through. For $m = 1$, $D_{2m} = \langle x, y \mid x = 1, y^2 = 1 \rangle = \{1, y\}$ is isomorphic to $\mathbb{Z}/2\mathbb{Z}$, and has two irreducible representations ψ_+ and ψ_-. Also, φ_0 is equivalent to $\psi_+ \oplus \psi_-$. For $m = 2$, $D_{2m} = \langle x, y \mid x^2 = 1, y^2 = 1, xy = yx \rangle = \{1, x, y, xy\}$ is isomorphic to $\mathbb{Z}/2\mathbb{Z} \oplus \mathbb{Z}/2\mathbb{Z}$, and has four irreducible representations ψ_{++}, ψ_{+-}, ψ_{-+}, and ψ_{--}. Also, φ_0 is equivalent to $\psi_{++} \oplus \psi_{+-}$, and φ_1 is equivalent to $\psi_{-+} \oplus \psi_{--}$.

Example 1.12. Consider the the group of unit quaternions

$$Q_8 = \{\pm\mathbf{1}, \pm\mathbf{i}, \pm\mathbf{j}, \pm\mathbf{k}\}$$

and note that this group has a quotient $Q_8/\{\pm 1\}$ that is isomorphic to D_4. Let $\pi : Q_8 \to D_4$ be the quotient map. Then for each 1-dimensional representation ψ of D_4, the composition $\psi\pi$ is a representation of Q_8. This gives four 1-dimensional representations. The group Q_8 also has a 2-dimensional representation ρ given by

$$\rho(\pm\mathbf{1}) = \begin{bmatrix} \pm 1 & 0 \\ 0 & \pm 1 \end{bmatrix}, \quad \rho(\pm\mathbf{i}) = \begin{bmatrix} \pm i & 0 \\ 0 & \mp i \end{bmatrix},$$

$$\rho(\pm\mathbf{j}) = \begin{bmatrix} 0 & \pm i \\ \pm i & 0 \end{bmatrix}, \quad \rho(\pm\mathbf{k}) = \begin{bmatrix} 0 & \mp 1 \\ \pm 1 & 0 \end{bmatrix}.$$

The proofs of the claims in Examples 1.10, 1.11, and 1.12 are all more or less difficult exercises in linear algebra. We shall carry out the more difficult of these later in the text. Note also that all occurrences of $\mathbf{C}$ in Example 1.10 could be replaced by $\mathbf{Q}[\exp(2\pi i/p)]$, and $\mathbf{C}$ in Example 1.11 could be replaced by $\mathbf{Q}[\exp(2\pi i/m)]$.

Our approach in this book will be via module theory, and so we recast Definition 1.1 in this language.

Definition 1.13. Let G be a group and $\mathbf{F}$ a field. An **F-representation** of G is a left $\mathbf{F}(G)$-module M.

Here $\mathbf{F}(G)$ denotes the group algebra of G over $\mathbf{F}$,

$$\mathbf{F}(G) = \left\{ \sum_{g \in G} a_g g \mid a_g \in \mathbf{F}, \text{only finitely many } a_g \neq 0 \right\}.$$

The equivalence of Definitions 1.1 and 1.13 should be clear.

Example 1.14. As an example we may take $M = \mathbf{F}(G)$. This is called the **regular representation** of G, and it plays a key role.

Now here is our agenda for this book. As we have seen above, an irreducible representation is indecomposable but not necessarily vice versa. However, it turns out that the situation in which both of these notions are equivalent is much easier than the general case. This arises when $R = \mathbf{F}(G)$ is "semisimple", so we first begin by studying semisimple rings and modules.

Next we want to study the question of the field of definition of representations (that is, the question of when $\mathbf{E}$-representations of G are defined over a subfield $\mathbf{F}$). In order to do so we study the behavior of modules under field extensions rather generally. Actually, we will think here about starting with $\mathbf{F}$ and taking extensions $\mathbf{E}$, and asking if new representations appear or old representations decompose. We will see that all the interesting behavior occurs between $\mathbf{E} = \mathbf{F}$ and $\mathbf{E} = \overline{\mathbf{F}}$, the algebraic closure of $\mathbf{F}$—once we pass $\overline{\mathbf{F}}$, nothing new happens.

We then study group representations over algebraically closed fields in the semisimple case. This arises (exactly) when G is finite and $\mathrm{char}(\mathbf{F}) = 0$ or is prime to the order of G. Thus, in addition to this case being simpler, it is really of central importance (including the case of complex representations of finite groups).

Next we study "induced representations". Induction is a method of constructing a representation of a group G from a representation of a subgroup of G. This turns out to be a powerful method and one that is of central importance in the theory.

From what we have just said, all of the complex representations of a finite group G are actually defined over the algebraic closure $\overline{\mathbf{Q}}$ of $\mathbf{Q}$. It was a conjecture of Schur, proved by Brauer, that they are in fact all defined over $\mathbf{Q}(\sqrt[u]{1})$, where u is exponent of G, i.e., the least common multiple of the orders of the elements of G. Our study of the semisimple case culminates in proving this result.

We then move on to the nonsemisimple case. As in the semisimple case, we begin by developing the requisite ring and module theory. We then present an introduction to the theory of nonsemisimple group representations.

Chapter 2

Semisimple Rings and Modules

In this chapter we develop the necessary ring and module theory for our applications to representation theory in the "semisimple" case. Actually, we will go a bit further than is strictly necessary, in order to prove some additional results that are interesting and useful in themselves.

2.1. Basic Notions

To begin with, we let R be an arbitrary ring. For our purposes, "ring" means ring with identity. Also "module" means left module, unless otherwise stated.

Definition 1.1. A nonzero R-module M is **simple**, or **irreducible**, if the only submodules of M are $\{0\}$ and M.

The following lemma is almost immediate, but turns out to be incredibly useful.

Lemma 1.2 (Schur's lemma).

(1) *Let M be a simple R-module. Then $\operatorname{End}_R(M) = \operatorname{Hom}_R(M, M)$ is a division ring.*

(2) *Let M and N be simple R-modules. Then $\operatorname{Hom}_R(M, N) \neq \{0\}$ if and only if M and N are isomorphic.*

(3) *Suppose that R is an $\mathbf{F}$-algebra, where $\mathbf{F}$ is algebraically closed, and M is a simple R-module with $\dim_{\mathbf{F}} M$ finite. Then $\mathrm{End}_R(M) = \mathbf{F}$, consisting entirely of homotheties (i.e., of multiplication by elements of $\mathbf{F}$).*

Proof.

(1) We need to show that any $\varphi \in \mathrm{End}_R(M), \varphi \neq 0$, is invertible. But $\varphi : M \to M$, so $\mathrm{Ker}(\varphi)$ and $\mathrm{Im}(\varphi)$ are submodules of M. Since M is simple and $\varphi \neq 0$, we must have $\mathrm{Ker}(\varphi) = \{0\}$ and $\mathrm{Im}(\varphi) = M$, so φ is an isomorphism and hence invertible.

(2) Assume $\varphi \in \mathrm{Hom}_R(M, N)$ is nonzero. Then $\mathrm{Ker}(\varphi)$ is a proper submodule of M, and $\mathrm{Im}(\varphi)$ is a nonzero submodule of N, so $\mathrm{Ker}(\varphi) = \{0\}$ and $\mathrm{Im}(\varphi) = N$ and φ is an isomorphism, so M and N are isomorphic. Conversely, if M and N are isomorphic, let $\varphi : M \to N$ be an isomorphism. Then $0 \neq \varphi \in \mathrm{Hom}_R(M, N)$.

(3) Let $\varphi \in \mathrm{End}_R(M)$. Then M is a finite-dimensional vector space over $\mathbf{F}$, and φ is an $\mathbf{F}$-linear map. Let $c(x)$ be the characteristic polynomial of φ. Since $\mathbf{F}$ is algebraically closed, $c(x)$ has a root $\lambda \in \mathbf{F}$, i.e., λ is an eigenvalue of φ. Then $\mathrm{Ker}(\lambda I - \varphi)$ is a nonzero submodule of M, so $\mathrm{Ker}(\lambda I - \varphi) = M$ as M is simple. In other words, $\varphi = \lambda I$ is just multiplication by $\lambda \in \mathbf{F}$.　□

Recall that a nonzero R-module M is called **cyclic** if it is generated by some nonzero element m of M. Also, for any R-module M and any $m \in M$, the **annihilator** of m is $\mathrm{Ann}(m) = \{r \in R \mid rm = 0\}$.

Proposition 1.3. *Let M be an R-module. Then M is simple if and only if M is cyclic and generated by any nonzero element of M.*

Proof. Suppose M is simple, and let $m \in M, m \neq 0$. Then $\langle m \rangle = Rm = \{rm \mid r \in R\}$ is the submodule of M generated by m. Since $\langle m \rangle \neq \{0\}$ and M is simple, $\langle m \rangle = M$. On the other hand, suppose M is not simple, and let M_1 be a nonzero proper submodule of M. Then for any $m \in M_1$, $\langle m \rangle \subseteq M_1 \subset M$, so m does not generate M.　□

Corollary 1.4. *Every simple R-module is isomorphic to a quotient of R.*

Proof. This is true for every cyclic R-module, as $\langle m \rangle \cong R/\mathrm{Ann}(m)$.　□

In conjunction with part (3) of Schur's lemma, it is worth making the following observation.

Corollary 1.5. *If R is a finite-dimensional $\mathbf{F}$-algebra and M is a simple R-module, then $\dim_{\mathbf{F}} M$ is finite.*

Proof.　□

Proposition 1.6. *A cyclic R-module $M = \langle m \rangle$ is simple if and only if $\mathrm{Ann}(m)$ is a maximal left ideal of R.*

Proof. $M \cong R/\operatorname{Ann}(m)$ so submodules of M are in one-to-one correspondence with ideals of R containing $\operatorname{Ann}(m)$. $\qquad\square$

Remark 1.7. In general, not every cyclic R-module is simple. For example, let $M = \mathbf{F}^2$ be an $\mathbf{F}(X)$-module, where X acts by multiplication by the matrix $\left[\begin{smallmatrix} 1 & 1 \\ 0 & 1 \end{smallmatrix}\right]$. Then M is cyclic but not simple, as $M = \mathbf{F}(X)\left[\begin{smallmatrix} 0 \\ 1 \end{smallmatrix}\right]$ but $\mathbf{F}(X)\left[\begin{smallmatrix} 1 \\ 0 \end{smallmatrix}\right]$ is a nonzero proper submodule.

Definition 1.8. A nonzero R-module M is **indecomposable** if $M = M_1 \oplus M_2$ implies $M_1 = \{0\}$ and $M_2 = M$ or vice versa.

Proposition 1.9. *If M is irreducible, then M is indecomposable.*

Proof. $\qquad\square$

Remark. The converse of Proposition 1.9 is in general false. The module M of Remark 1.7 is indecomposable but not irreducible.

Definition 1.10. An R-module M is **semisimple** if M is a direct sum of simple R-modules.

2.2. Structure Theorems

In this section we will analyze semisimple R-modules over an arbitrary ring R, and then analyze the structure of semisimple rings R (which have the property that *every* R-module is semisimple). As we will see later, $R = \mathbf{F}(G)$ is semisimple if and only if G is finite and $\operatorname{char}(\mathbf{F})$ is zero or prime to the order of G. We will be applying the theory developed here to the representation theory of G in this (important) case.

We choose once and for all a set of representatives $\{M_\alpha\}_{\alpha \in A}$ of the isomorphism classes of simple R-modules.

If M is an R-module and s is a positive integer, we let sM denote the direct sum $M \oplus \cdots \oplus M$ (s summands). More generally, if Γ is any index set we let ΓM denote the R-module $\Gamma M = \bigoplus_{\gamma \in \Gamma} M_\gamma$, where $M_\gamma = M$ for all $\gamma \in \Gamma$. Of course if $|\Gamma| = s < \infty$, then $\Gamma M = sM$, and we will prefer the latter notation.

This notation is convenient for describing semisimple modules as direct sums of simple R-modules. If P is a semisimple R-module, then

$$(2.1) \qquad P = \bigoplus_{i \in I} P_i,$$

where P_i is simple for each $i \in I$. If we collect all the simple modules in equation (2.1) that are isomorphic, then we obtain

$$(2.2) \qquad P \cong \bigoplus_{\alpha \in A} (\Gamma_\alpha M_\alpha).$$

Equation (2.2) is said to be a **simple factorization** of the semisimple module P.

More precisely, let $f : \bigoplus_{\alpha \in A} (\Gamma_\alpha M_\alpha) \to P$ be an isomorphism. For fixed α, $f(\Gamma_\alpha M_\alpha) \subseteq P$ is called the M_α-**isotypic** component of P, and we denote it by P^α. The following results show that P^α is independent of the choice of isomorphism f, and also provide a uniqueness result for the simple factorization.

Theorem 2.1.

(1) *Let P be a semisimple R-module with simple factorization*

$$P \cong \bigoplus_{\alpha \in A} (\Gamma_\alpha M_\alpha).$$

Then, for each $\alpha \in A$, P^α, the M_α-isotypic component of P, is well defined (i.e., independent of the choice of f).

(2) *Let Q be a semisimple R-module with simple factorization*

$$Q \cong \bigoplus_{\alpha \in A} (\Lambda_\alpha M_\alpha).$$

Let $\varphi : P \to Q$ be an isomorphism. Then, for each $\alpha \in A$, $\varphi : P^\alpha \to Q^\alpha$ is an isomorphism.

Proof. We begin by proving part (2) for *an* M_α-isotypic component of P and of Q. Then applying part (2) with $Q = P$ and $\varphi = 1_P$ shows that P^α is well defined (i.e., that part (1) holds), which then enables us to conclude part (2) for *the* M_α-isotypic components.

Let $f : \bigoplus_{\alpha \in A} (\Gamma_\alpha M_\alpha) \to P$ and $g : \bigoplus_{\alpha \in A} (\Lambda_\alpha M_\alpha) \to Q$ be isomorphisms. For fixed α, let $P^\alpha = f(\Gamma_\alpha M_\alpha)$, $Q^\alpha = g(\Lambda_\alpha M_\alpha)$, and furthermore let $\overline{P}^\alpha = \bigoplus_{\alpha' \neq \alpha} P^\alpha$, $\overline{Q}^\alpha = \bigoplus_{\alpha' \neq \alpha} Q^\alpha$. Thus $P = P^\alpha \oplus \overline{P}^\alpha$ and $Q = Q^\alpha \oplus \overline{Q}^\alpha$.

Since φ is an isomorphism, $\varphi|P^\alpha$ is an injection. For fixed α, let $\iota : P^\alpha \to P$ be the natural inclusion and $\pi : Q \to \overline{Q}^\alpha$ be the natural projection. Then the composition $\pi \varphi \iota : P^\alpha \to \overline{Q}^\alpha$ is the zero map, by Schur's lemma, as P^α is the direct sum of simple R-modules all of which are isomorphic to M_α, and $\overline{Q}^\alpha$ is the direct sum of simple R-modules none of which are isomorphic to M_α. Hence $\varphi(P^\alpha) \subseteq Q^\alpha$ for each α, so $\varphi|P^\alpha : P^\alpha \to Q^\alpha$ is an injection for each α. Furthermore, since φ is an isomorphism, φ is a surjection, and hence $\varphi|P^\alpha : P^\alpha \to Q^\alpha$ is a surjection for each α. Thus, $\varphi|P^\alpha : P^\alpha \to Q^\alpha$ is an isomorphism for each α. $\qquad\square$

Theorem 2.2.

(1) *Let P be a semisimple R-module with simple factorizations*

$$P \cong \bigoplus_{\alpha \in A} (\Gamma_\alpha M_\alpha) \cong \bigoplus_{\alpha \in A} (\Lambda_\alpha M_\alpha).$$

Then for each α, $|\Gamma_\alpha| = |\Lambda_\alpha| \in \{0, 1, 2, \dots\} \cup \{\infty\}$.

(2) *Let P and Q be semisimple R-modules with simple factorizations*

$$P \cong \bigoplus_{\alpha \in A} (\Gamma_\alpha M_\alpha)$$

and

$$Q \cong \bigoplus_{\alpha \in A} (\Lambda_\alpha M_\alpha).$$

Suppose that P and Q are isomorphic. Then $|\Gamma_\alpha| < \infty$ if and only if $|\Lambda_\alpha| < \infty$, in which case $|\Gamma_\alpha| = |\Lambda_\alpha|$.

Proof. Again we prove part (2) for some pair of simple factorizations, and then apply (2) with $Q = P$ to conclude that part (1) holds, and then that part (2) holds for any pair of simple factorizations.

Let $\varphi : P \to Q$ be an isomorphism. Fix an α. If Γ_α and Λ_α are both infinite, there is nothing to prove. Suppose not. For simplicity we assume that both $n_\Gamma = |\Gamma_\alpha|$ and $n_\Lambda = |\Lambda_\alpha|$ are finite. We prove (2) by induction on n_Γ. As $P^\alpha \neq \{0\}$ if and only if $n_\Gamma \neq 0$, and $Q^\alpha \neq \{0\}$ if and only if $n_\Lambda \neq 0$, we see that $n_\Gamma = 0$ implies $n_\Lambda = 0$. Now let $n_\Gamma = n > 0$ and assume the theorem is true for all pairs of R modules with $n_\Gamma = n - 1$. Let $\gamma_0 \in \Gamma_\alpha$ be arbitrary, and choose $\lambda_0 \in \Lambda_\alpha$ so that the composition $\pi_{\lambda_0} \varphi \iota_{\gamma_0} : M_\alpha \to M_\alpha$ is nonzero, and hence an isomorphism (by Schur's lemma, as M_α is simple), where ι_{γ_0} is the inclusion onto the γ_0-th summand and π_{λ_0} is the projection onto the λ_0-th summand. Then $P^\alpha/\iota_{\gamma_0}(M_\alpha)$ is isomorphic to $Q^\alpha/\iota_{\lambda_0}(M_\alpha)$. But $P^\alpha/\iota_{\gamma_0}(M_\alpha)$ is isomorphic to $\bigoplus_{\alpha \in \Gamma'_\alpha} (\Gamma_\alpha M_\alpha)$ and $Q^\alpha/\iota_{\lambda_0}(M_\alpha)$ is isomorphic to $\bigoplus_{\alpha \in \Lambda'_\alpha} (\Lambda_\alpha M_\alpha)$, where $\Gamma'_\alpha = \Gamma_\alpha \setminus \{\gamma_0\}$ and $\Lambda'_\alpha = \Lambda_\alpha \setminus \{\lambda_0\}$. These sets have cardinality $n_\Gamma - 1$ and $n_\Lambda - 1$ respectively, so by induction $n_\Gamma - 1 = n_\Lambda - 1$ and hence $n_\Gamma = n_\Lambda$, as required. (The proof is easily modified to handle the case that one of $|\Gamma_\alpha|$ and $|\Lambda_\alpha|$ is infinite, by inducting on whichever one of $|\Gamma_\alpha|$ and $|\Lambda_\alpha|$ that is finite.) $\qquad\square$

Remark 2.3. While it is true that the M^α-isotypic components of a semisimple R-module are well defined, it is not necessarily true that the decomposition of these components into simple summands is. For example, let $R = \mathbf{F}$ be a field, so there is only a single isomorphism class of simple R-modules, namely that of a 1-dimensional $\mathbf{F}$-vector space. Let P be a vector space over $\mathbf{F}$ of dimension s. Then for *any* choice of basis $\{b_1, \ldots, b_s\}$ of P, we obtain a direct sum decomposition

$$P \cong Rb_1 \oplus \cdots \oplus Rb_s.$$

Definition 2.4. Let M be a semisimple R-module with simple factorization

$$M \cong \bigoplus_{\alpha \in A} (\Gamma_\alpha M_\alpha).$$

Then $\mathcal{T}(M) = \{\Gamma_\alpha M_\alpha\}$ (a set with multiplicities) is the **type** of M and $\ell(M) = \sum |\Gamma_\alpha|$ is the **length** of M.

Note that, by Theorem 2.2, $\mathcal{T}(M)$ and $\ell(M)$ are well defined. We will see later that Theorem 2.2 and Definition 2.4 have appropriate generalizations to nonsemisimple modules.

We now present some characterizations of semisimple modules.

Lemma 2.5. *Let M be an R-module that is a sum of simple submodules $\{M_i\}_{i \in I}$, and let N be an arbitrary submodule of M. Then there is a subset $J \subseteq I$ such that*

$$M = N \oplus \left(\bigoplus_{i \in J} M_i \right).$$

Proof. The proof is an application of Zorn's lemma. For $P \subseteq I$, we let $M_P \subseteq M$ be the sum of the simple submodules $\{M_i\}_{i \in P}$. Now let

$$S = \left\{ P \subseteq I \mid M_P = \bigoplus_{i \in P} M_i \quad \text{and} \quad M_P \cap N = \{0\} \right\}.$$

(By definition, M_P is the sum of the modules $\{M_i\}_{i \in P}$. In the definition of S we are requiring that it be the *direct* sum of these modules. This subtle difference is key to the proof.) Partially order S by inclusion and let $\mathcal{C} = \{P_k\}_{k \in K}$ be an arbitrary chain in S. If $P = \bigcup_{k \in K} P_k$, we claim that $P \in S$. Suppose that $P \notin S$. Since it is clear that $M_P \cap N = \{0\}$, we must have that $M_P \neq \bigoplus_{i \in P} M_i$. Then there is some $p_0 \in P$ such that $M_{p_0} \cap M_{P'} \neq \{0\}$, where $P' = P \setminus \{p_0\}$. Suppose that $0 \neq x \in M_{p_0} \cap M_{P'}$. Then we may write

$$(2.3) \qquad\qquad x = x_{p_1} + \cdots + x_{p_j},$$

where $x_{p_i} \neq 0 \in M_{p_i}$ for $\{p_1, \ldots, p_j\} \subseteq P'$. Since $\mathcal{C}$ is a chain, there is an index $k \in K$ such that $\{p_0, p_1, \ldots, p_j\} \subseteq P_k$. Equation (2.3) shows that $M_{P_k} \neq \bigoplus_{i \in P_k} M_i$, which contradicts the fact that $P_k \in \mathcal{S}$. Therefore, we must have $P \in \mathcal{S}$, and Zorn's lemma applies to conclude that $\mathcal{S}$ has a maximal element J. To complete the proof of the lemma we need only prove the following claim:

Claim. $M = N + M_J = N \oplus \left(\bigoplus_{i \in J} M_i \right)$.

Note that the second equality in the claim is true by the definition of $\mathcal{S}$, so it remains to prove that the first equality is true. Suppose not. Then there is an index $i_0 \in I$ such that $M_{i_0} \not\subseteq N + M_J$. Hence $M_{i_0} \cap (N + M_J)$ is a proper submodule of M_{i_0}. Since M_{i_0} is simple, we must then have $M_{i_0} \cap (N + M_J) = \{0\}$, and this implies that $N \cap (M_{i_0} + M_J) = \{0\}$. (If $n = m_{i_0} + m_J$ with $n \in N$, $m_{i_0} \in M_{i_0}$, and $m_J \in M_J$, then $m_{i_0} = n - m_J$, so $n - m_J = 0$, i.e., $n = m_J$, and then $n = 0$ as $N \cap M_J = \{0\}$.) Also, $M_{i_0} \cap M_J \subseteq M_{i_0} \cap (N + M_J) = \{0\}$ so $M_{i_0} \cap M_J = \{0\}$. Therefore, $\{i_0\} \cup J \in \mathcal{S}$, contradicting the maximality of J, and hence proving the claim. $\qquad\square$

Corollary 2.6. *If an R-module M is a sum of simple submodules, then M is semisimple.*

Proof. Take $N = \{0\}$ in Lemma 2.5. $\qquad\square$

Theorem 2.7. *If M is an R-module, then the following are equivalent:*

(1) *M is a semisimple R-module.*

(2) *Every submodule of M is complemented (i.e., for every submodule N of M, there is a submodule N' of M with $M = N \oplus N'$).*

(3) *Every submodule of M is a sum of simple R-modules.*

Proof. $(1) \Rightarrow (2)$ follows from Lemma 2.5, and $(3) \Rightarrow (1)$ is immediate from Corollary 2.6. It remains to prove $(2) \Rightarrow (3)$.

Let M_1 be a submodule of M. First we observe that every submodule of M_1 is complemented in M_1. To see this, suppose that N is any submodule of M_1. Then N is complemented in M, so there is a submodule N' of M such that $N \oplus N' = M$, i.e., $N + N' = M$ and $N \cap N' = \{0\}$. But then $N + (N' \cap M_1) = M_1$ and $N \cap (N' \cap M_1) = \{0\}$, i.e., $N \oplus (N' \cap M_1) = M_1$, and hence N is complemented in M_1.

Next we claim that every nonzero submodule M_2 of M contains a nonzero simple submodule. To see this, let $m \in M_2$, $m \neq 0$. Then $Rm \subseteq M_2$ and, furthermore, $R/\operatorname{Ann}(m) \cong Rm$, where $\operatorname{Ann}(m) = \{a \in R \mid am = 0\}$ is a left ideal of R. A simple Zorn's lemma argument shows that there is a maximal left ideal I of R containing $\mathrm{Ann}(m)$. Then Im is a maximal submodule of Rm. By the last paragraph, Im is a complemented submodule of Rm, so there is a submodule N of Rm with $N \oplus Im = Rm$, and since Im is a maximal submodule of Rm, it follows that the submodule N is simple. Thus, we have produced a simple submodule N of M_2.

Now consider an arbitrary submodule N of M, and let $N_1 \subseteq N$ be the sum of all the simple submodules of N. We claim that $N_1 = N$. N_1 is complemented in N, so we may write $N = N_1 \oplus N_2$. If $N_2 \neq \{0\}$, then N_2 has a nonzero simple submodule N', and, since $N' \subseteq N$, by the definition of N_1, we have that $N' \subseteq N_1$. Hence $N' \subseteq N_1 \cap N_2$. But $N_1 \cap N_2 = \{0\}$ as N_2 is a complement of N_1. This contradiction shows that $N_2 = \{0\}$, i.e., $N = N_1$, and the proof is complete. $\qquad\square$

Corollary 2.8. *Sums, submodules, and quotient modules of semisimple modules are semisimple.*

Proof. Sums: This follows immediately from Corollary 2.6.

Submodules: Any submodule of a semisimple module satisfies condition (3) of Theorem 2.7.

Quotient modules: If M is a semisimple module, $N \subseteq M$ a submodule, and $Q = M/N$, then N has a complement N' in M, i.e., $M = N \oplus N'$. But then $Q \cong N'$, so Q is isomorphic to a submodule of M, and hence is semisimple. $\qquad\square$

Corollary 2.9. *Let M be a semisimple R-module and let $N \subseteq M$ be a submodule. Then N is irreducible (simple) if and only if N is indecomposable.*

Proof. Since every irreducible module is indecomposable, we need only show that if N is not irreducible, then N is not indecomposable. Let N_1 be a nontrivial proper submodule of N. Then N is semisimple by Corollary 2.8, so N_1 has a complement by Theorem 2.7, and so N is not indecomposable. $\square$

Remark 2.10. The fact that every submodule of a semisimple R-module M is complemented is equivalent to the statement that whenever M is a semisimple R-module, every short exact sequence

$$0 \longrightarrow N \longrightarrow M \longrightarrow K \longrightarrow 0$$

of R-modules splits.

Definition 2.11. A ring R is called **semisimple** if R is semisimple as a left R-module.

Remark. The proper terminology should be "left semisimple", with an analogous definition of "right semisimple", but we shall see below that the two notions coincide.

Theorem 2.12. *Let R be a ring. The following are equivalent:*

(1) *R is a semisimple ring.*

(2) *Every R-module is semisimple.*

(3) *Every R-module is projective.*

Proof. (1) $\Rightarrow$ (2). Let M be an R-module. Then M has a free presentation

$$0 \longrightarrow K \longrightarrow F \longrightarrow M \longrightarrow 0$$

so that M is a quotient of the free R-module F. Since F is a direct sum of copies of R and R is assumed to be semisimple, it follows that F is semisimple, and hence, by Corollary 2.8, M is as well.

(2) $\Rightarrow$ (3). Assume that every R-module is semisimple, and let P be an arbitrary R-module. Suppose that

$$(2.4) \qquad 0 \longrightarrow K \longrightarrow M \longrightarrow P \longrightarrow 0$$

is a short exact sequence. Since M is an R-module, by assumption it is semisimple, and then Remark 2.10 implies that sequence (2.4) is split exact. Since (2.4) is an arbitrary short exact sequence with P on the right, it follows that P is projective.

(3) $\Rightarrow$ (1). Let M be an arbitrary submodule of R (i.e., an arbitrary left ideal). Then we have a short exact sequence

$$0 \longrightarrow M \longrightarrow R \longrightarrow R/M \longrightarrow 0.$$

Since all R-modules are assumed projective, we have that R/M is projective, and hence this sequence splits. Therefore, $R = M \oplus N$ for some submodule $N \subseteq R$, which is isomorphic (as an R-module) to R/M. Then, by Theorem 2.7, R is semisimple. $\square$

Corollary 2.13. *Let R be a semisimple ring and let M be an R-module. Then M is irreducible (simple) if and only if M is indecomposable.*

Proof. $\square$

Theorem 2.14. *Let R be a semisimple ring. Then every simple R-module is isomorphic to a submodule of R.*

Proof. Let N be a simple R-module, and let $R = \bigoplus_{i \in I} M_i$ with each M_i simple. We must show that at least one of the simple R-modules M_i is isomorphic to N. If this is not the case, then

$$\operatorname{Hom}_R(R, N) \cong \operatorname{Hom}_R\left(\bigoplus_{i \in I} M_i, N\right) \cong \prod_{i \in I} \operatorname{Hom}_R(M_i, N) = \{0\},$$

where the last equality holds because $\operatorname{Hom}_R(M_i, N) = \{0\}$ if M_i is not isomorphic to N (Schur's lemma). But $\operatorname{Hom}_R(R, N) \cong N \neq \{0\}$, and this contradiction shows that we must have N isomorphic to one of the simple submodules M_i of R. $\qquad\square$

Corollary 2.15. *Let R be a semisimple ring.*

(1) *There are only finitely many isomorphism classes of simple R-modules.*

(2) *If $\{M_\alpha\}_{\alpha \in A}$ is a set of representatives of the isomorphism classes of simple R-modules, then*

$$R \cong \bigoplus_{\alpha \in A} n_\alpha M_\alpha$$

with each n_α a positive integer.

Proof. Since R is semisimple, we may write

$$R = \bigoplus_{i \in I} M_i,$$

where each M_i is simple. We will show that I is finite, and then the fact that there are only finitely many isomorphism classes and that each appears with finite multiplicity is immediate from Theorem 2.14.

Consider the identity element $1 \in R$. By the definition of direct sum, we have

$$1 = \sum_{i \in I} m_i$$

for some elements $m_i \in M_i$, with all but finitely many m_i equal to zero. Of course, each M_i is a left R-submodule of R, i.e., a left ideal.

Now suppose that I is infinite. Then there is an $i_0 \in I$ for which $m_{i_0} = 0$. Let m be any nonzero element of M_{i_0}. Then

$$m = m \cdot 1 = m\left(\sum_{i \in I} m_i\right) = \sum_{i \in I \setminus \{i_0\}} m m_i$$

so $m \in \bigoplus_{i \in I \setminus \{i_0\}} M_i$.

Thus,

$$m \in M_{i_0} \cap \left(\bigoplus_{I \in I \setminus \{i_0\}} M_i\right) = \{0\},$$

by the definition of direct sum again, which is a contradiction. Hence, I is finite.

Finally, each n_α is nonzero, again by Theorem 2.14. $\qquad\square$

We now come to the basic structure theorem for semisimple rings.

Theorem 2.16 (Wedderburn). *Let R be a semisimple ring. Then there is a finite set A, a collection of integers $\{n_\alpha \mid \alpha \in A\}$, and division rings $\{D_\alpha \mid \alpha \in A\}$ such that*

$$R \cong \bigoplus_{\alpha \in A} \mathrm{End}_{D_\alpha}(D_\alpha^{n_\alpha}).$$

Proof. First, a general observation: For a ring R, we let R^{op} denote the ring with the same underlying set as R, but with multiplication given by $r \cdot s = sr$, for $r, s \in R$. Then $R^{op} \cong \mathrm{End}_R(R)$ with the isomorphism given by $r \mapsto \varphi_r : R \to R$ with $\varphi_r(s) = sr$ for $s \in R$.

Now, to the proof of this theorem: By Corollary 2.15, we may write R as a finite direct sum:

$$R \cong \bigoplus_{\alpha \in A} n_\alpha M_\alpha.$$

By Schur's lemma, $\mathrm{End}_R(M_\alpha)$ is a division algebra, which we denote by E_α, for each $\alpha \in A$. Then R is anti-isomorphic to R^{op}, and (again using Schur's lemma to obtain the third isomorphism below)

$$
\begin{aligned}
R^{op} &\cong \mathrm{End}_R(R) \\
&\cong \mathrm{Hom}_R\Big(\bigoplus_{\alpha \in A} n_\alpha M_\alpha, \bigoplus_{\alpha \in A} n_\alpha M_\alpha \Big) \\
&\cong \bigoplus_{\alpha \in A} \mathrm{Hom}_R(n_\alpha M_\alpha, n_\alpha M_\alpha) \\
&\cong \prod_{i=1}^{n_\alpha} \bigoplus_{i=1}^{n_\alpha} \mathrm{Hom}_R(M_\alpha, M_\alpha) \\
&\cong \prod_{i=1}^{n_\alpha} \bigoplus_{i=1}^{n_\alpha} E_\alpha \\
&\cong \prod_{i=1}^{n_\alpha} \bigoplus_{i=1}^{n_\alpha} \mathrm{Hom}_{E_\alpha}(E_\alpha, E_\alpha) \\
&\cong \mathrm{Hom}_{E_\alpha}(n_\alpha E_\alpha, n_\alpha E_\alpha) \\
&= \mathrm{End}_{E_\alpha}(E_\alpha^{n_\alpha}).
\end{aligned}
$$

Setting $D_\alpha = E_\alpha^{op}$, we complete the proof by observing that $\mathrm{End}_{E_\alpha}(E_\alpha^{n_\alpha})$ is anti-isomorphic to $\mathrm{End}_{D_\alpha}(D_\alpha^{n_\alpha})$. $\square$

Remark. Note that by [AW, Corollary 4.3.9], $\mathrm{End}_D(D^n)$ is isomorphic to $\mathrm{Mat}_n(D^{op})$, where for a ring R, $\mathrm{Mat}_n(R)$ denotes the ring of n-by-n matrices with entries in R. Thus, Wedderburn's theorem is often stated as: *Every semisimple ring is isomorphic to a finite direct sum of matrix rings over division rings.*

Lemma 2.17. *Let D be a division ring and n a positive integer. Then $R = \operatorname{End}_D(D^n)$ is semisimple as a left R-module and also as a right R-module. Furthermore, R is semisimple as a left D-module and as a right D-module.*

Proof. Write $D^n = D_1 \oplus D_2 \oplus \cdots \oplus D_n$ where $D_i = D$. Let

$$M_i = \left\{ f \in \operatorname{End}_D(D^n) : \operatorname{Ker}(f) \supseteq \bigoplus_{k \neq i} D_k \right\},$$

$$N_j = \{ f \in \operatorname{End}_D(D^n) : Im(f) \subseteq D_j \},$$

and let

$$P_{ij} = M_i \cap N_j.$$

Note that $P_{ij} \cong D$. Then

$$\operatorname{End}_D(D^n) \cong M_1 \oplus \cdots \oplus M_n$$

as a left R-module, and

$$\operatorname{End}_D(D^n) \cong N_1 \oplus \cdots \oplus N_n$$

as a right R-module. We leave it to the reader to check that each M_i (resp. N_j) is a simple left (resp. right) R-module. Also,

$$\operatorname{End}_D(D^n) \cong \bigoplus P_{ij}$$

as a left (resp. right) D-module, and each P_{ij} is certainly simple (on either side). $\qquad\square$

Corollary 2.18. *A ring R is semisimple as a left R-module if and only if it is semisimple as a right R-module.*

Proof. This follows immediately from Theorem 2.16 and Lemma 2.17. $\quad\square$

Observe that R is a simple left R-module (resp. right R-module) if and only if R has no nontrivial proper left (resp. right) ideals, which is the case if and only if R is a division algebra. Thus, to define simplicity of R in this way would bring nothing new. Instead we make the following definition:

Definition 2.19. A ring R with identity is **simple** if it has no nontrivial proper two-sided ideals.

Corollary 2.20. *Let D be a division ring and n a positive integer. Then $\mathrm{End}_D(D^n)$ is a simple ring that is semisimple as a left $\mathrm{End}_D(D^n)$-module.*

Conversely, if R is a simple ring that is semisimple as a left R-module, or, equivalently, as a right R-module, then

$$R \cong \mathrm{End}_D(D^n)$$

for some division ring D and positive integer n.

Proof. We leave it to the reader to check that $\mathrm{End}_D(D^n)$ is simple (compare [AW, Theorem 2.2.26 and Corollary 2.2.27]), and then the first part of the corollary follows from Lemma 2.17. Conversely, if R is semisimple, we have the decomposition given by Wedderburn's theorem (Theorem 2.16), and then the condition of simplicity forces $k = 1$. $\square$

The language suggests that a simple ring should be semisimple. This is almost the case.

Proposition 2.21. *Let R be a simple ring. Then R is semisimple if and only if R has a minimal left ideal.*

Proof. Let I be a minimal left ideal. Note that I is then a simple R-module. We claim that for any $r \in R$, either $Ir = \{0\}$ or Ir is also a minimal left ideal, and hence a simple R-module. If $Ir = \{0\}$ there is nothing to prove. Otherwise, let s be any nonzero element of Ir, so $s = ir$ for some i in I. We need only show that $Rs = Ir$. Clearly $\{0\} \neq Ri \subseteq RI = I$, so $Ri = I$ by minimality. Then $Rs = R(ir) = (Ri)r = Ir$, as required.

Now consider IR. This is a two-sided ideal in R. But R is simple, so $IR = R$. In other words, R is the sum of the simple R-modules Ir for each $r \in R$. But then R is semisimple by Corollary 2.6.

Conversely, if R is simple and semisimple, then $R \cong \mathrm{End}_D(D^n)$ by Corollary 2.20. But then, as in the proof of Lemma 2.17, R contains a simple (left) R-module, i.e., a minimal left ideal. $\square$

With the help of Wedderburn's theorem, we can obtain more precise information than in Corollary 2.15 in the case that R is a finite-dimensional algebra over **F**.

Theorem 2.22. *Let R be a semisimple ring that is finite dimensional as an **F**-algebra. Write*

$$(2.5) \qquad R \cong \bigoplus_{\alpha \in A} n_\alpha M_\alpha,$$

where $\{M_\alpha\}$ are representatives of the distinct isomorphism classes of simple R-modules, and set $E_\alpha = \mathrm{End}_R(M_\alpha)$, a division ring. Then

(1) $\dim_{\mathbf{F}} \operatorname{Center}(R) = \sum_{\alpha \in A} \dim_{\mathbf{F}} \operatorname{Center}(E_\alpha)$

(2) $\dim_{\mathbf{F}} M_\alpha = n_\alpha \dim_{\mathbf{F}} E_\alpha$

(3) $\dim_{\mathbf{F}} R = \sum_{\alpha \in A} n_\alpha^2 \dim_{\mathbf{F}} E_\alpha$.

Proof. From the proof of Wedderburn's theorem we see that

$$(2.6) \qquad\qquad R \cong \bigoplus_{\alpha \in A} \operatorname{Mat}_{n_\alpha}(E_\alpha).$$

Taking the dimension over $\mathbf{F}$ of both sides of (2.6) gives (3).

Applying $\operatorname{Hom}_R(\_, M_\alpha)$ to both sides of (2.5), and using Schur's lemma, we have

$$M_\alpha \cong \operatorname{Hom}_R(R, M_\alpha) \cong \operatorname{Hom}_R\Big(\bigoplus_{\beta \in A} n_\beta M_\beta, M_\alpha\Big)$$

$$\cong \operatorname{Hom}_R(n_\alpha M_\alpha, M_\alpha) \cong \prod_{i=1}^{n_\alpha} \operatorname{Hom}_R(M_\alpha, M_\alpha) \cong n_\alpha E_\alpha,$$

yielding (2).

Finally, taking the dimension over $\mathbf{F}$ of the centers of both sides of (2.6), and observing that the center of $\operatorname{Mat}_{n_\alpha}(E_\alpha)$ consists precisely of scalar matrices with entries in the center of E_α, we obtain (1). $\qquad\square$

We usually regard Theorem 2.22 as a way of determining the integers $|A|$ and $\{n_\alpha\}$. In general, this involves investigating the division rings E_α. However, there is a (very) important case where the situation is much easier. First, a lemma.

Lemma 2.23. *Let $\mathbf{F}$ be an algebraically closed field and E a division ring that is a finite-dimensional $\mathbf{F}$-algebra. Then $E = \mathbf{F}$.*

Proof. For any fixed element e of $\mathbf{E}$, let $m_e : E \to E$ be left multiplication by e, $m_e(e') = ee'$. Then m_e is an $\mathbf{F}$-linear map on the finite-dimensional $\mathbf{F}$-linear vector space E, and E is algebraically closed, so m_e has an eigenvalue $\lambda \in \mathbf{F}$ and an eigenvector e'. Then $(e - \lambda)e' = 0$, with $e' \neq 0$. But E is a division ring, so $e = \lambda$. $\qquad\square$

Corollary 2.24. *Let R be a semisimple ring that is finite dimensional as an $\mathbf{F}$-algebra, where $\mathbf{F}$ is algebraically closed. Write*

$$R \cong \bigoplus_{\alpha \in A} n_\alpha M_\alpha,$$

where $\{M_\alpha\}$ are the distinct isomorphism classes of simple R-modules. Then

(1) $|A| = \dim_{\mathbf{F}} \text{Center}(R)$

(2) $n_\alpha = \dim_{\mathbf{F}} M_\alpha$

(3) $\dim_{\mathbf{F}} R = \sum_{\alpha \in A} n_\alpha^2$.

Proof. $\qquad\qquad\qquad\qquad\qquad\qquad\qquad\qquad\qquad\qquad\qquad\qquad\qquad\qquad\square$

2.3. Idempotents and Blocks

In this section we will see how to use idempotents in order to decompose "acceptable" rings into blocks. At the end of this section we will specialize to semisimple rings.

First, however, some words of caution are in order.

Let $\{R_i\}_{i \in I}$ be rings. Then $R = \bigoplus_{i \in I} R_i$ with coordinatewise multiplication satisfies all the axioms for a ring except possibly the existence of a unit. More precisely, R will have a unit if and only if I is finite. In this case, if 1_i is the unit of R_i, $1 = \sum_i 1_i$ is the unit of R. In this situation we will say that $R = \bigoplus R_i$ as rings. Note that each R_i is a two-sided ideal in R (but not a subring, as the identity of R_i is not the identity of R).

Also, suppose $R = I_1 \oplus I_2$ as R-modules (or equivalently, as left ideals), and let $1 = i_1 + i_2$. Then i_1 and i_2 need not be units in I_1 and I_2, respectively, and I_1 and I_2 need not have units. Furthermore, it is *not* necessarily the case that $R = I_1 \oplus I_2$ as rings (with or without unit). For example, let $R = GL_2(\mathbf{F}) = \{[\begin{smallmatrix} a & b \\ c & d \end{smallmatrix}]\}$, $I_1 = \{[\begin{smallmatrix} a & 0 \\ c & 0 \end{smallmatrix}]\}$, $I_2 = \{[\begin{smallmatrix} 0 & b \\ 0 & d \end{smallmatrix}]\}$. Then $R = I_1 \oplus I_2$ as ideals but not as rings. Also, $[\begin{smallmatrix} 1 & 0 \\ 0 & 1 \end{smallmatrix}] = [\begin{smallmatrix} 1 & 0 \\ 0 & 0 \end{smallmatrix}] + [\begin{smallmatrix} 0 & 0 \\ 0 & 1 \end{smallmatrix}]$ but $[\begin{smallmatrix} 1 & 0 \\ 0 & 0 \end{smallmatrix}]$ and $[\begin{smallmatrix} 0 & 0 \\ 0 & 1 \end{smallmatrix}]$ are not units.

Definition 3.1. Let R be a ring. An element e of R is an **idempotent** if $e^2 = e$. Two idempotents e and f are **orthogonal** if $ef = fe = 0$. A finite set of idempotents $\{e_i\}$ that are pairwise mutually orthogonal and with $\sum e_i = 1$ are called **complementary**.

Definition 3.2. An idempotent e of R is **central** if $e \in \text{Center}(R)$. An idempotent e is **primitive** if $e = f_1 + f_2$ with f_2 and f_2 orthogonal idempotents implies $f_1 = e$ and $f_2 = 0$ or vice versa. A central idempotent e is **central primitive** if $e = f_1 + f_2$ with f_1 and f_2 orthogonal central idempotents implies $f_1 = e$ and $f_2 = 0$ or vice versa.

Example 3.3.

(1) If R has no zero divisors, the only idempotents of R are 0 and 1. If R is a simple ring, the only central idempotents of R are 0 and 1.

(2) If e is an idempotent of R, then $f = 1 - e$ is also an idempotent, and $\{e, f\}$ are complementary.

(3) If $R = \operatorname{End}(V)$, V a vector space, then the idempotents of R are precisely the projections $\pi : V \to W$ onto subspaces of V. Such an idempotent is primitive if and only if $\dim(W) = 1$.

(4) If G is a finite group and $\operatorname{char}(\mathbf{F}) = 0$ or is prime to $|G|$, then $\frac{1}{|G|}\sum_{g \in G} g$ is an idempotent in $\mathbf{F}(G)$. This idempotent is central. More generally, if H is any finite subgroup of a group G and $\operatorname{char}(\mathbf{F}) = 0$ or is prime to $|H|$, then $\frac{1}{|H|}\sum_{h \in H} h$ is an idempotent in $\mathbf{F}(G)$. This idempotent is central if and only if H is a normal subgroup of G.

Lemma 3.4.

(1) *If $e \neq 0$ is a central idempotent, then Re is a ring with unit e.*

(2) *If $e \neq 0$ is an idempotent and Re is a ring, then e is the unit.*

Proof.

(1) Clear.

(2) Let re have unit u, and write $u = ve$. Then $e = ue = (ve)e = ve^2 = ve = u$. $\qquad\square$

Lemma 3.5.

(1) *For any set $\{e_i\}$ of complementary nonzero idempotents, $R = \bigoplus Re_i$ as R-modules.*

(2) *If $\{e_i\}$ are also central, then $R = \bigoplus Re_i$ as rings.*

Proof.

(1) For any $r \in R$, $r = r1 = r(\sum e_i) = \sum re_i$ so R is the sum of the ideals Re_i. Suppose $0 = r_1 e_1 + \cdots + r_k e_k$. Then $0 = 0e_i = (r_1 e_1 + \cdots + r_k e_k)e_i = r_i e_i^2 = r_i e_i$ for all i, so R is the direct sum.

(2) We compute: $(\sum_i a_i e_i)(\sum_j b_j e_j) = \sum_{i,j} a_i e_i b_j e_j = \sum_{i,j} a_i b_j e_i e_j = \sum_i a_i b_i e_i^2 = \sum_i (a_i e_i)(b_i e_i)$. $\qquad\square$

Lemma 3.6.

(1) *Suppose $R = \bigoplus R_i$, a finite direct sum, as R-modules. Write $1 = \sum e_i$, $e_i \in R_i$. Then $\{e_i\}$ are complementary idempotents.*

(2) *Suppose $R = \bigoplus R_i$, a finite direct sum, as rings. Write $1 = \sum e_i$, $e_i \in R_i$. Then $\{e_i\}$ are central.*

Proof.

(1) Let $\pi_i : R \to R_i$ be the projection. Then $e_i = \pi_i(1)$ and so $e_i e_j = e_i \pi_j(1) = \pi_j(e_i) = e_i$ if $j = i$ and 0 if $j \neq i$.

(2) For $r \in R$, write $r = \sum r_i$. Then $r = r1 = (\sum r_i)(\sum e_i) = \sum r_i e_i$, and $r = 1r = (\sum e_i)(\sum r_i) = \sum e_i r_i$, so $e_i r_i = r_i e_i$ for all i, and hence $e_i r = r e_i$ for all i. $\qquad\square$

Corollary 3.7. *Let $R = \bigoplus R_i$ as R-modules, and let $\pi : R \to R_i$ be the projection. Then $e_i = \pi_i(1)$ is an idempotent and $R_i = R e_i$. If $R = \bigoplus R_i$ as rings, then $e_i = \pi_i(1)$ is a central idempotent.*

Proof. $\qquad\square$

We say that R is **indecomposable (as a ring)** if it has no nontrivial direct sum decomposition (as rings).

For the remainder of this section $\{e_i\}$ will denote idempotents and $\{e_\alpha\}$ will denote central idempotents.

Lemma 3.8. *Let R be a ring.*

(1) *The idempotent $e \neq 0$ of R is primitive if and only if the R-module Re is indecomposable.*

(2) *The central idempotent $e \neq 0$ of R is central primitive if and only if the ring Re is indecomposable.*

Proof.

(1) Suppose $Re = I_1 \oplus I_2$. Let $\pi : R \to Re$ and $\pi_i : R \to I_i$ be the projections. Then $\pi = \pi_1 \oplus \pi_2$ so $e = \pi(1) = \pi_1(1) + \pi_2(1) = e_1 + e_2$ is not primitive. Conversely, if $e = e_1 + e_2$ with e_1 and e_2 orthogonal, and $I_i = Re_i$, then $Re = I_1 \oplus I_2$ as follows: $I_1 \cap I_2 = \{0\}$ as if $r_1 e_1 = r_2 e_2$, then $0 = e_1 e_2 = r_1 e_1 e_2 = r_2 e_2 e_2 = r_2 e_2 = r_1 e_1$; $Re \subseteq I_1 \oplus I_2$ as $re = re_1 + re_2$; and $Re \supseteq I_1 + I_2$ as $(r_1 e_1 + r_2 e_2)e = (r_1 e_1 + r_2 e_2)(e_1 + e_2) = r_1 e_1 + r_2 e_2$.

(2) The proof is similar. $\qquad\square$

Definition 3.9. A ring R is **acceptable** if it is a finite direct sum of indecomposable left ideals.

Remark. This is not standard terminology.

Lemma 3.10. *R is acceptable if and only if it has a complementary set of primitive idempotents.*

Proof. If $R = \bigoplus R_i$, a finite direct sum of indecomposable left ideals, let $\pi : R \to R_i$ and let $e_i = \pi_i(1)$. Then $\{e_i\}$ is a complementary set of primitive idempotents, by Lemmas 3.6 and 3.8. Conversely, if $\{e_i\}$ is a complementary set of primitive idempotents, (this set being finite by definition), $R = \bigoplus Re_i$ and each Re_i is indecomposable, also by Lemma 3.8. $\qquad\square$

So far R has been arbitrary. However, for the rest of this section we will assume that R is acceptable.

Lemma 3.11. *Let R be acceptable. Then R has only finitely many central idempotents.*

Proof. Let $\{e_i\}$ be a set of complementary primitive idempotents, and let f be a central idempotent. Then $e_i = e_i f + e_i(1 - f)$, each of which is an idempotent as f is central. Since e_i is primitive, either $e_i f = 0$ or $e_i f = e_i$. Then $f = 1 \cdot f = (\sum e_i)f = \sum a_i e_i$ with $a_i = 0$ or 1, so there are only finitely many possibilities. $\qquad\square$

Definition 3.12. Let $R = \bigoplus R_\alpha$ as rings, with each R_α an indecomposable ring. Then each R_α is a **block**, and $R = \bigoplus R_\alpha$ is a **block decomposition**.

Corollary 3.13. *Let $R = \bigoplus R_\alpha$ be a block decomposition, and let $\pi_\alpha : R \to R_\alpha$ be the projection. Then $e_\alpha = \pi_\alpha(1)$ is a central primitive idempotent. Conversely, if $\{e_\alpha\}$ is a complementary set of central primitive idempotents, then $R = \bigoplus Re_\alpha$ is a block decomposition.*

Proof. This follows directly from Lemma 3.5, Lemma 3.6, Corollary 3.7, and Lemma 3.8. $\qquad\square$

Lemma 3.14. *Let $\{e_\alpha\}$ and $\{f_\beta\}$ be two sets of complementary nonzero central-primitive idempotents of R. Then $\{e_\alpha\} = \{f_\beta\}$.*

Proof. For each β_0, $f_{\beta_0} = 1 f_{\beta_0} = (\sum e_\alpha)f_{\beta_0} = \sum e_\alpha f_{\beta_0}$, so $e_{\alpha_0} f_{\beta_0} \neq 0$ for some α_0. But then $R f_{\beta_0} = R e_{\alpha_0} f_{\beta_0} \oplus R(1 - e_{\alpha_0})f_{\beta_0}$ as rings. But $R f_{\beta_0}$ is indecomposable, so $R e_{\alpha_0} f_{\beta_0} = R f_{\beta_0}$. Similarly, $R e_{\alpha_0} f_{\beta_0} = R f_{\beta_0} e_{\alpha_0} = R e_{\alpha_0}$. But now $R f_{\beta_0} = R e_{\alpha_0}$ has a unique identity element, which is $f_{\beta_0} = e_{\alpha_0}$. $\square$

Theorem 3.15. *Let R be an acceptable ring. Then R has a block decomposition. Furthermore, if $R = \bigoplus R_\alpha = \bigoplus S_\beta$ are two block decompositions of R, then $\{R_\alpha\} = \{S_\beta\}$.*

Proof. Existence follows directly from Lemma 3.10 and Corollary 3.13. As for uniqueness, $R = \bigoplus R_\alpha = \bigoplus R\pi_\alpha(1)$ where $\pi_\alpha : R \to R_\alpha$, and $\pi_\alpha(1)$ is a central primitive idempotent. Similarly, $R = \bigoplus S_\beta = \bigoplus R\pi'_\beta(1)$. But then $\{\pi_\alpha(1)\} = \{\pi'_\beta(1)\}$ by Lemma 3.14. $\qquad\square$

Remark. The theorem states: *An acceptable ring has a unique block decomposition.* Note also that unique means unique, not just unique up to isomorphism.

Remark. While there is uniqueness for blocks, there is only uniqueness up to isomorphism for decompositions $R = \bigoplus R_i$ as R-modules with the R_i indecomposable. To see this, take the case $R = \text{End}(V)$, V a vector space. Then corresponding to *any* direct sum decomposition $V = \bigoplus W_i$ of

V into 1-dimensional subspaces, we obtain a decomposition $R = \bigoplus R_i$ as R-modules with the R_i indecomposable R-modules. Indeed, in this case the R_i are simple R-modules.

For the remainder of this section we will assume that R is semisimple.

Lemma 3.16. *Let R be a semisimple ring. Then R is acceptable.*

Proof. Write
$$R = \bigoplus_{i \in I} M_i$$

with the M_i simple left ideals. Then, by Corollary 2.15, this is a finite sum, so R is acceptable. $\square$

Lemma 3.17. *Let R be a semisimple ring and $e \neq 0$ a central idempotent of R. Then the ring Re is semisimple.*

Proof. As R is semisimple, Re is semisimple as an R-module. But note that $I \subseteq Re$ is an R-module if and only if $I \subseteq Re$ is an Re-module. (Clearly I is invariant under Re if it is invariant under R. Conversely, suppose I is invariant under Re. Let $i \in I$ and $r \in R$ be arbitrary. As $I \subseteq Re$, $i = je$ for some j. Then $ri = rje = rje^2 = reje = (re)i \in I$, so I is also invariant under R.) $\square$

Write
$$R \cong \bigoplus_{\alpha \in A} n_\alpha M_\alpha,$$

a simple factorization. Then, in the notation of the proof of Lemma 3.16, $M_i = Re_i$ and $\{e_i\}_{i \in I}$ is a complementary set of primitive idempotents. For each $\alpha \in A$, let

$$I_\alpha = \{i \in I \mid M_i \text{ is isomorphic to } M_\alpha\}$$

(so $|I_\alpha| = n_\alpha$), and set

$$e_\alpha = \sum_{i \in I_\alpha} e_i.$$

Then $R = \bigoplus Re_\alpha$ as R-modules.

Proposition 3.18. $R = \bigoplus Re_\alpha$ *as rings, and each Re_α is a simple ring.*

Proof. First we claim that if $N_i = Rf_i$, $i = 1, 2$, are simple R-modules, with f_i idempotents, that are not isomorphic, then $f_1 r f_2 = 0$ for every $r \in R$. To see this, define $\varphi : N_1 \to N_2$ by $\varphi(n_1) = n_1 r f_2$. By Schur's lemma, $\varphi = 0$, so in particular $\varphi(f_1) = f_1 r f_2 = 0$.

Now let $r = \sum_{\alpha \in A,\ i \in I_\alpha} r_{\alpha,i} e_{\alpha,i}$ and $s = \sum_{\beta \in B,\ i \in I_\beta} s_{\beta,j} e_{\beta,j}$. Then

$$rs = \left(\sum r_{\alpha,i} e_{\alpha,i}\right)\left(\sum s_{\beta,j} e_{\beta,j}\right) = \sum\sum r_{\alpha,i} e_{\alpha i} s_{\beta,j} e_{\beta,j}$$
$$= \sum r_{\alpha,i} e_{\alpha,i} s_{\alpha,j} e_{\alpha,j}$$

so the multiplication is coordinatewise and hence $R = \bigoplus Re_\alpha$ as rings. Now we must show each Re_α is simple. But $Re_\alpha \cong \bigoplus n_\alpha M_\alpha$ as R-modules, and hence as Re_α-modules, so, as in the proof of Wedderburn's theorem, Re_α is anti-isomorphic to $\mathrm{End}_{Re_\alpha}(n_\alpha M_\alpha)$, which is a simple ring. $\qquad\square$

Remark. Let $J \subseteq I$ be a subset with $\#(J \cap I_\alpha) = 1$ for each $\alpha \in A$. Then $\{e_J\}_{j \in J}$ are primitive idempotents (necessarily orthogonal) and $\{Re_j\}_{j \in J}$ is a complete set of representatives of isomorphism classes of simple R-modules, so there are $|J| = |A|$ of them. Conversely, if we find $|A|$ primitive idempotents giving pairwise nonisomorphic simple R-modules, we have found all of them, up to isomorphism. Note that $|A|$ is the number of summands in the block decomposition $R = \bigoplus_{\alpha \in A} R_\alpha = \bigoplus_{\alpha \in A} Re_\alpha$.

In fact, one method of constructing group representations is to write down idempotents in $\mathbf{F}(G)$. Conversely, if we have constructed group representations in some other way, we can use that knowledge to write down idempotents.

Remark. Comparing Wedderburn's theorem, we see that if $R = \bigoplus_{\alpha \in A} R_\alpha$ as rings, with each R_α a simple ring, then $R_\alpha \cong \mathrm{End}_{D_\alpha}(D_\alpha^{n_\alpha})$, where $n_\alpha = |I_\alpha|$ and $D_\alpha = \mathrm{End}_R(M_\alpha)$.

For future use we observe the following:

Lemma 3.19. *Let R be a semisimple ring that is an $\mathbf{F}$-algebra, and suppose that $\dim_{\mathbf{F}} \mathrm{Center}(R) = t$ is finite. If $\{e_{\alpha_1}, \dots, e_{\alpha_t}\}$ is a complementary set of central idempotents, then each e_{α_i} is central primitive, and so $R = \bigoplus Re_{\alpha_i}$ is a block decomposition.*

Proof. $\{e_{\alpha_1}, \dots, e_{\alpha_t}\}$ is linearly independent over $\mathbf{F}$, as $0 = \sum_j e_{\alpha_j} = e_{\alpha_i}(\sum c_j e_{\alpha_j}) = c_i e_{\alpha_i}^2 = c_i e_{\alpha_i}$ implies $c_i = 0$ for each i. Thus this is a linearly independent set of $t = \dim_{\mathbf{F}} \mathrm{Center}(R)$ elements of $\mathrm{Center}(R)$, so it is a basis of $\mathrm{Center}(R)$. Thus, if f is any central element, $f = \sum c_i e_{\alpha_i}$ for some $c_i \in \mathbf{F}$. Then $f^2 = \sum c_i^2 e_{\alpha_i}$, so f an idempotent implies $c_i = 0$ or 1 for each i, which easily implies that each e_{α_i} is central primitive. $\qquad\square$

2.4. Behavior under Field Extensions

Now we consider the case where the ring R is an $\mathbf{F}$-algebra, for some field $\mathbf{F}$. Let $\mathbf{E}$ be an extension field of $\mathbf{F}$, and set $R' = \mathbf{E} \otimes_{\mathbf{F}} R$. In this section we will investigate the question: How do R-modules and R'-modules compare? The meaning of this rather vague question will become clear in a moment.

Definition 4.1. Let M be an R-module, and set $M' = \mathbf{E} \otimes_{\mathbf{F}} M$, an R'-module. Then M and M' are said to **correspond**. Also, if M' is an R'-module that is isomorphic to $\mathbf{E} \otimes_{\mathbf{F}} M$ for some R-module M, then M' is said to be **defined over $\mathbf{F}$**.

Notation. We use the following notation throughout this section:

(1) $R' = \mathbf{E} \otimes_{\mathbf{F}} R$.

(2) M always denotes an R-module.

(3) M' always denotes an R'-module.

(4) $\mathbf{E}(M)$ denotes $\mathbf{E} \otimes_{\mathbf{F}} M$.

As we pass from $\mathbf{F}$ to ever larger fields $\mathbf{E}$, we may expect that essentially new R'-modules (i.e., R'-modules not corresponding to R-modules) will appear. Because of this, modules that are irreducible over R may correspond to modules that are not irreducible over R'. Thus, things may change, and we will be interested in seeing how they do, or whether they stay the same.

Key to our investigations will be two technical results—Schur's lemma (Lemma 1.2) and the following theorem.

Theorem 4.2. *Let M_1 and M_2 be R-modules that are finite dimensional over $\mathbf{F}$ and let $M_1' = \mathbf{E}(M_1)$ and $M_2' = \mathbf{E}(M_2)$ be the corresponding R'-modules. Then*

$$\operatorname{Hom}_{R'}(M_1', M_2') \cong \mathbf{E} \otimes_{\mathbf{F}} \operatorname{Hom}_R(M_1, M_2).$$

Proof. Let $\{e_i\}$ be a basis for $\mathbf{E}$ as an $\mathbf{F}$-vector space. Then the conclusion of the theorem is exactly the claim that any $\varphi \in \operatorname{Hom}_{R'}(M_1', M_2')$ can be written uniquely as $\varphi = \sum e_i \varphi_i$, where $\varphi_i \in \operatorname{Hom}_R(M_1, M_2)$.

In order to keep track of the action of R (or R'), let us write $\sigma_i(r)(m)$ for the action of R on M_i, $i = 1, 2$, and similarly for $\sigma_i'(r)(m)$.

Pick bases for M_1 and M_2 (over $\mathbf{F}$); then there are also bases for M_1' and M_2' (over $\mathbf{E}$). Let $[\ ]$ denote matrices with respect to these bases. Note $[\sigma_i'(r)] = [\sigma_i(r)]$, $i = 1, 2$.

Now let $\varphi : M_1' \to M_2'$ be an arbitrary R'-module map. Then $[\sigma_1'(r)]\,[\varphi]$ $= [\varphi]\,[\sigma_2'(r)]$ for every $r \in R'$. Since $\{e_i\}$ is a basis for $\mathbf{E}$ over $\mathbf{F}$, we may write uniquely

$$[\varphi] = \sum e_i A_i,$$

where A_i are matrices with entries in $\mathbf{F}$. Let $\varphi_i : M_1 \to M_2$ be the linear transformation defined by $[\varphi_i] = A_i$. Then $\varphi = \sum e_i \varphi_i$, and this expression is clearly unique. It remains only to show that each $\varphi_i : M_i \to M_2$ is an R-module map. But

$$\begin{aligned}
0 &= [\sigma_1'(r)][\varphi] - [\varphi][\sigma_2'(r)] \\
&= [\sigma_1'(r)]\Big(\sum_i e_i A_i\Big) - \Big(\sum_i e_i A_i\Big)[\sigma_2'(r)] \\
&= \sum_i e_i\big([\sigma_1(r)]A_i - A_i[\sigma_2(r)]\big) \\
&= \sum_i e_i\big([\sigma_1(r)][\varphi_i] - [\varphi_i][\sigma_2(r)]\big),
\end{aligned}$$

and since the $\{e_i\}$ are linearly independent over $\mathbf{F}$, we have $[\sigma_1(r)][\varphi_i] = [\varphi_i][\sigma_2(r)]$, as required. $\square$

Corollary 4.3. *Under the hypotheses of Theorem 4.2,*

$$\dim_{\mathbf{E}} \mathrm{Hom}_{R'}(M_1', M_2') = \dim_{\mathbf{F}} \mathrm{Hom}_R(M_1, M_2).$$

Proof. $\square$

Henceforth we assume that R and R' are semisimple, for every extension field $\mathbf{E}$ of $\mathbf{F}$, and that $\dim_{\mathbf{F}} R = \dim_{\mathbf{E}} R'$ is finite. (As we will see, this will be true in case $R = \mathbf{F}(G)$, G a finite group, F a field of characteristic 0 or prime to the order of G.) Note then, by Theorem 2.14, that every irreducible R-module (resp. irreducible R'-module) is finite dimensional as a vector space over $\mathbf{F}$ (resp. over $\mathbf{E}$).

Corollary 4.4. *Let M_1 and M_2 be two distinct irreducible R-modules. Then $M_1' = \mathbf{E}(M_1)$ and $M_2' = \mathbf{E}(M_2)$ have no common irreducible sub-R'-module.*

Proof. By Schur's lemma, $\mathrm{Hom}_R(M_1, M_2) = \{0\}$ so by the last corollary $\mathrm{Hom}_{R'}(M_1', M_2') = \{0\}$, which is equivalent to the conclusion of the corollary, again by Schur's lemma. $\square$

Corollary 4.5.

(1) *Let M_1 and M_2 be R-modules. Let $M_1' = \mathbf{E}(M_1)$ and $M_2' = \mathbf{E}(M_2)$. Then M_1 and M_2 are isomorphic if and only if M_1' and M_2' are isomorphic for some, and hence for every, $\mathbf{E} \supseteq \mathbf{F}$.*

(2) *Let M_2 be a finite sum of irreducible R-modules. Let $M_1' = \mathbf{E}(M_1)$ and $M_2' = \mathbf{E}(M_2)$. Suppose that $M_1' \cong M_2' \oplus M_3'$. Then $M_1 \cong M_2 \oplus M_3$ for some R-module M_3 with $M_3' = \mathbf{E}(M_3)$.*

Proof.

(1) Only one direction needs proof. First suppose that M_1 and M_2 are irreducible. Suppose that M_1' and M_2' are isomorphic for some $\mathbf{E} \supseteq \mathbf{F}$. Then $\{0\} \neq \operatorname{Hom}_{R'}(M_1', M_2') \cong \mathbf{E} \otimes_{\mathbf{F}} \operatorname{Hom}_R(M_1, M_2)$ so $\{0\} \neq \operatorname{Hom}_R(M_1, M_2)$ and M_1 and M_2 are isomorphic by Schur's lemma. The general case then follows from Corollary 4.4.

(2) We prove this in case M_2 is irreducible. The general case then follows by induction.

Now M_2' a summand of M_1' implies that $\operatorname{Hom}_{R'}(M_2', M_1') \neq \{0\}$, which implies that $\operatorname{Hom}_R(M_2, M_1) \neq \{0\}$, which implies that M_2 is a summand of M_1, i.e., that $M_1 \cong M_2 \oplus M_3$ for some R-module M_3. Then

$$M_2' \oplus M_3' \cong M_1' = \mathbf{E}(M_1) = \mathbf{E}(M_2 \oplus M_3) = M_2' \oplus \mathbf{E}(M_3)$$

so M_3' and $\mathbf{E}(M_3)$ are isomorphic. (M_2' is a finite sum of irreducible R'-modules, as $\dim_{\mathbf{E}} M_2' = \dim_{\mathbf{F}} M_2 \leq \dim_{\mathbf{F}} R$ is finite, and then this follows from the "unique factorization" Theorem 2.2.) $\qquad\square$

Notation. We let $t(R)$ be the number of distinct isomorphism classes of irreducible R-modules.

Note that $t(R)$ is finite by Corollary 2.15.

Corollary 4.6. *If $\mathbf{E} \supseteq \mathbf{F}$, then $t(R') \geq t(R)$.*

Proof. If $M_1, \dots, M_{t(R)}$ are the irreducible R-modules, and $M_i' = \mathbf{E}(M_i)$, then each of $M_1', \dots, M_{t(R)}'$ contains at least one irreducible R'-module, and, by Corollary 4.4, they are all distinct. $\qquad\square$

Definition 4.7. An irreducible R-module M is **absolutely irreducible** if $M' = \mathbf{E}(M)$ is irreducible for every $\mathbf{E} \supseteq \mathbf{F}$.

Notation. We denote the algebraic closure of $\mathbf{F}$ by $\overline{\mathbf{F}}$, and set $\overline{R} = \overline{\mathbf{F}} \otimes_{\mathbf{F}} R$, $\overline{M} = \overline{\mathbf{F}} \otimes_{\mathbf{F}} M = \overline{\mathbf{F}}(M)$.

A priori, we might have to examine infinitely many fields to decide whether an irreducible representation is absolutely irreducible. But we do not—we can tell by looking at $\mathbf{F}$ (or $\overline{\mathbf{F}}$).

Corollary 4.8. *Let $\overline{M} = \mathbf{F}(M)$. The following are equivalent:*

(1) *M is absolutely irreducible.*

(2) $\mathrm{Hom}_R(M, M) \cong \mathbf{F}$.

(3) $\mathrm{Hom}_{\overline{R}}(\overline{M}, \overline{M}) \cong \overline{\mathbf{F}}$.

(4) *$\overline{M}$ is irreducible.*

Proof. M absolutely irreducible $\Rightarrow \overline{M}$ irreducible $\Rightarrow \mathrm{Hom}_{\overline{R}}(\overline{M}, \overline{M}) \cong \overline{\mathbf{F}}$ (by Schur's lemma) $\Rightarrow \mathrm{Hom}_R(M, M) \cong \mathbf{F}$ (by Corollary 4.3).

M not absolutely irreducible $\Rightarrow M' = \mathbf{E}(M)$ not irreducible for some $\mathbf{E} \supseteq \mathbf{F} \Rightarrow \mathrm{Hom}_{\overline{\mathbf{E} \otimes_\mathbf{E} R'}}(\overline{\mathbf{E}} \otimes_\mathbf{E} M', \overline{\mathbf{E}} \otimes_\mathbf{E} M') \not\cong \overline{\mathbf{E}} \Rightarrow \mathrm{Hom}_{\overline{R}}(\overline{M}, \overline{M}) \not\cong \overline{\mathbf{F}}$ (as $\overline{\mathbf{E}} \supseteq \overline{\mathbf{F}}$) $\Rightarrow \overline{M}$ not irreducible and $\mathrm{Hom}_R(M, M) \not\cong \mathbf{F}$. $\square$

Lemma 4.9. *Let $\mathbf{E} \supseteq \mathbf{F}$ be algebraically closed. Then there is a one-to-one correspondence between irreducible $\overline{R}$-modules and irreducible $\mathbf{E} \otimes_\mathbf{F} R$-modules given by $\overline{M} \mapsto \mathbf{E} \otimes_{\overline{\mathbf{F}}} \overline{M}$. In particular, $t(\mathbf{E} \otimes_\mathbf{F} R) = t(\overline{R})$.*

Proof. (First note that $\mathbf{E} \otimes_\mathbf{F} R = \mathbf{E} \otimes_{\overline{\mathbf{F}}} \overline{R}$.) Recall we have a simple factorization of $\overline{R}$,

$$\overline{R} = \bigoplus_{\alpha \in A} n_\alpha \overline{M}_\alpha,$$

where $\{\overline{M}_\alpha\}$ are representatives of the distinct isomorphism classes of irreducible $\overline{R}$-modules. But then

$$\mathbf{E} \otimes_{\overline{\mathbf{F}}} \overline{R} \cong \bigoplus_{\alpha \in A} n_\alpha (\mathbf{E} \otimes_{\overline{\mathbf{F}}} \overline{M}_\alpha).$$

Now each $\overline{M}_\alpha$ is irreducible, hence absolutely irreducible (as $\mathbf{F}$ is algebraically closed), so each $\mathbf{E} \otimes_{\overline{\mathbf{F}}} \overline{M}_\alpha$ is irreducible. As $\{\overline{M}_\alpha\}$ are distinct, so are $\{\mathbf{E} \otimes_{\overline{\mathbf{F}}} \overline{M}_\alpha\}$ (by Corollary 4.4), so we have a simple factorization of $\mathbf{E} \otimes_{\overline{\mathbf{F}}} \overline{R}$ and $t(\mathbf{E} \otimes_\mathbf{F} R) = |A| = t(\overline{R})$. $\square$

As we have seen, as we pass from $\mathbf{F}$ to $\mathbf{E}$, an irreducible module M might correspond to a module M' that is no longer irreducible. Consider each of the irreducible submodules of M' and pass to an even larger field, etc. Sooner or later this process will cease producing anything new, and it is that which we investigate now.

Definition 4.10. A field $\mathbf{E} \supseteq \mathbf{F}$ is a **splitting field** for R if every irreducible R'-module is absolutely irreducible.

Corollary 4.11. *The following are equivalent:*

(1) *$\mathbf{E}$ is a splitting field for R.*

(2) $\operatorname{Hom}_{R'}(M', M') = \mathbf{E}$ *for every irreducible R'-module M'.*

Proof. This is just the equivalence of conditions (1) and (2) in Corollary 4.8. $\square$

The following corollary tells us splitting fields always exist.

Corollary 4.12. *Any algebraically closed field $\mathbf{E} \supseteq \mathbf{F}$ (and in particular $\mathbf{E} = \overline{\mathbf{F}}$) is a splitting field for R.*

Proof. Take $\mathbf{E}$ to be algebraically closed in Corollary 4.11. $\square$

But this is really only half the answer. We have that all irreducible R-modules *defined over a splitting field* are absolutely irreducible, but we would also like to know that all irreducible R-modules *are* defined over a splitting field.

We have to formulate this condition carefully. We cannot just mean that if $\mathbf{D}_0 \supseteq \mathbf{F}$ is *any* field and $\hat{M}$ any $\mathbf{D}_0 \otimes_{\mathbf{F}} R$-module, then $\hat{M}$ is defined over $\mathbf{E}$, i.e., $\hat{M} = \mathbf{D}_0 \otimes_{\mathbf{E}} M'$ for some R'-module M', as that would implicitly require that $\mathbf{D}_0 \supseteq \mathbf{E}$. Rather, we observe that if $\mathbf{D}_0 \supseteq \mathbf{F}$ is any field, $\hat{M}$ is any $(\mathbf{D}_0 \otimes_{\mathbf{F}} R)$-module, and $\mathbf{D}$ is any field containing $\mathbf{D}_0$, then $\mathbf{D} \otimes_{\mathbf{D}_0} \hat{M}$ is a $(\mathbf{D} \otimes_{\mathbf{F}} R)$-module (defined over $\mathbf{D}_0$). Now if we choose $\mathbf{D}$ to be any field containing both $\mathbf{D}_0$ and $\mathbf{E}$, it makes sense to require that $\mathbf{D} \otimes_{\mathbf{D}_0} \hat{M}$ be defined over $\mathbf{E}$, and that is what we do. But now we see that it is entirely superfluous to mention $\mathbf{D}_0$ in the first place, as we could simply take $\mathbf{D}_0 = \mathbf{D}$. Also, it is clear that we may restrict our attention to irreducible modules. This leads us to the following formulation, where we first show that we may restrict our attention to the case $\mathbf{D} = \overline{\mathbf{E}}$.

Definition 4.13. Let $\mathbf{E} \supseteq \mathbf{F}$. Then $\mathbf{E}$ is a **universal field of definition** for R if for any field $\mathbf{D} \supseteq \mathbf{E}$, every irreducible $(\mathbf{D} \otimes_{\mathbf{F}} R)$-module is defined over $\mathbf{E}$.

Lemma 4.14. *If every irreducible $(\overline{\mathbf{E}} \otimes_{\mathbf{F}} R)$-module is defined over $\mathbf{E}$, then $\mathbf{E}$ is a universal field of definition for R.*

Proof. Let $\mathbf{D}$ be any field containing $\mathbf{E}$, and let $\hat{M}$ be any irreducible $(\mathbf{D} \otimes_{\mathbf{F}} R)$-module. Then $\overline{\mathbf{D}} \otimes_{\mathbf{D}} \hat{M}$ is a $(\overline{\mathbf{D}} \otimes_{\mathbf{F}} R)$-module. But $\overline{\mathbf{D}} \supseteq \overline{\mathbf{E}}$, and both are algebraically closed, so, by Lemma 4.9, there is a one-to-one correspondence between $(\overline{\mathbf{D}} \otimes_{\mathbf{F}} R)$-modules and $(\overline{\mathbf{E}} \otimes_{\mathbf{F}} R)$-modules, i.e., $\overline{\mathbf{D}} \otimes_{\mathbf{D}} \hat{M}$ is defined over $\overline{\mathbf{E}}$, and hence over $\mathbf{E}$, by hypothesis (since every representation is a sum of irreducibles). In other words, $\overline{\mathbf{D}} \otimes_{\mathbf{D}} \hat{M} \cong \overline{\mathbf{D}} \otimes_{\mathbf{E}} M' \cong \overline{\mathbf{D}} \otimes_{\mathbf{D}} (\mathbf{D} \otimes_{\mathbf{E}} M')$

for some R'-module M'. But this implies $\hat{M} \cong \mathbf{D} \otimes_{\mathbf{E}} M'$, so $\hat{M}$ is defined over $\mathbf{E}$. $\square$

Theorem 4.15. *Let $\mathbf{E} \supseteq \mathbf{F}$. The following are equivalent:*

(1) $\mathbf{E}$ *is a universal field of definition for R.*

(2) $\mathbf{E}$ *is a splitting field for R.*

Proof. Let $\overline{\mathbf{E}}$ be the algebraic closure of $\mathbf{E}$, and set $\hat{R} = \overline{\mathbf{E}} \oplus_{\mathbf{E}} R'$. We have simple factorizations

$$\hat{R} \cong \bigoplus_{\beta \in B} \hat{n}_\beta \hat{M}_\beta$$

and

$$R' \cong \bigoplus_{\alpha \in A} n'_\alpha M'_\alpha.$$

The second of these yields

$$\hat{R} \cong \bigoplus_{\alpha \in A} n'_\alpha (\overline{\mathbf{E}} \otimes_{\mathbf{E}} M'_\alpha),$$

and we prove the theorem by comparing these two factorizations of $\hat{R}$.

Regardless of the nature of $\mathbf{E}$, since each $\hat{M}_\beta$ is irreducible, it is isomorphic to a submodule of some $\overline{\mathbf{E}} \otimes_{\mathbf{E}} M'_\alpha$, and in fact of exactly one $\overline{\mathbf{E}} \otimes_{\mathbf{E}} M'_\alpha$ by Corollary 4.4. Thus there is a well defined map $\varphi : B \to A$ given by $\varphi(\beta) = \alpha$ if $\hat{M}_\beta$ is isomorphic to a submodule of $\overline{\mathbf{E}} \otimes_{\mathbf{E}} M'_\alpha$, and φ is a surjection as every $\overline{\mathbf{E}} \otimes_{\mathbf{E}} M'_\alpha$ is a direct sum of irreducible $\hat{R}$-modules.

Now suppose $\mathbf{E}$ is a universal field of definition for R. Then each $\hat{M}_\beta$ is defined over R', i.e., is of the form $\overline{\mathbf{E}} \otimes_{\mathbf{E}} M'$ for some R'-module M'. Since $\hat{M}_\beta$ is irreducible, M' must be too, so $\hat{M}$ is isomorphic to some $\hat{M}_\beta$, and then M' is isomorphic to $M'_{\varphi(\beta)}$. Thus each $M'_{\varphi(\beta)}$ has the property that $\overline{\mathbf{E}} \otimes_{\mathbf{E}} M'_{\varphi(\beta)}$ is irreducible, i.e., each $M'_{\varphi(\beta)}$ is absolutely irreducible, and so each M'_α is absolutely irreducible (as φ is a surjection) and $\mathbf{E}$ is a splitting field for R.

On the other hand, suppose that $\mathbf{E}$ is a splitting field for R. Then for each M'_α, $\overline{\mathbf{E}} \otimes_{\mathbf{E}} M'_\alpha$ is irreducible, and these are distinct, again by Corollary 4.4. Thus $\bigoplus_{\alpha \in A} n_\alpha (\overline{\mathbf{E}} \otimes_{\mathbf{E}} M'_\alpha)$ is also a simple factorization of $\hat{R}$, and these two simple factorizations must agree, i.e., each $\hat{M}_\beta$ is isomorphic to $\overline{\mathbf{E}} \otimes_{\mathbf{E}} M'_\alpha$

for some α, or in other words, each $\widehat{M}_\beta$ is defined over $\mathbf{E}$, so $\mathbf{E}$ is a universal field of definition for R. $\qquad\square$

We have seen that $\overline{\mathbf{F}}$ is always a splitting field for R. In fact, a much smaller field will do.

Theorem 4.16. *R has a splitting field that is a finite extension of $\mathbf{F}$.*

Proof. Recall that we are assuming that $\dim_{\mathbf{F}} R$ is finite. Let $\{r_1, \ldots, r_d\}$ be a basis for R over $\mathbf{F}$, and note that, for any $\mathbf{E} \supseteq \mathbf{F}$, $\{r_1, \ldots, r_d\}$ is a basis for R' over $\mathbf{E}$, where we identify $1 \otimes r_i$ with r_i.

Since every irreducible R'-module M' is a summand of R', we have that $\dim_{\mathbf{E}} M'$ is finite.

Consider a simple factorization of $\overline{R}$,

$$\overline{R} \cong \bigoplus_{\beta \in B} \overline{n}_\beta \overline{M}_\beta.$$

Then, choosing a basis for $\overline{M}_\beta$ over $\overline{\mathbf{F}}$, we have a homomorphism $\sigma_\beta : \overline{R} \to \mathrm{Mat}_m(\overline{\mathbf{F}})$, $m = \dim_{\overline{\mathbf{F}}} \overline{M}_\beta$.

Consider $\{\sigma_\beta(r_i), i = 1, \ldots, d\}$. This is a finite collection of matrices, each of whose entries is an element of $\overline{\mathbf{F}}$. Let $\mathbf{E}_\beta$ be the field generated by all of the entries of all of these matrices. Then $\mathbf{E}_\beta$ is a finite extension of $\mathbf{F}$, and $\overline{M}_\beta$ is defined over $\mathbf{E}_\beta$. But now the set B is finite, so $\{\mathbf{E}_\beta \mid \beta \in B\}$ generates a finite extension of $\mathbf{F}$, which is a universal field of definition, and hence a splitting field, for R. $\qquad\square$

We make further observation. Let $\mathbf{E} \supseteq \mathbf{F}$. A priori, it is certainly possible that some (irreducible) R'-module M' is *not* defined over R, but some multiple mM' *is*, and indeed this occurs. In this situation we can say the following.

Proposition 4.17. *Let M' be an irreducible R'-module and suppose that some multiple of M' is defined over R. Let d be the smallest positive integer such that dM' is defined over R, i.e., $dM' = \mathbf{E} \otimes_{\mathbf{F}} M = \mathbf{E}(M)$ for some R-module M. Then M is irreducible. Furthermore, mM' is defined over R if and only if m is divisible by d.*

Proof. Let $dM' = \mathbf{E}(M)$. If M is reducible, $M = M_1 \oplus M_2$, then $dM' = \mathbf{E}(M_1) \oplus \mathbf{E}(M_2)$, so, as M' is irreducible, $\mathbf{E}(M_1) \cong d_1 M'$ and $\mathbf{E}(M_2) \cong d_2 M'$ for some d_1 and d_2 with $d_1 + d_2 = d$, by uniqueness of simple factorization (Theorem 2.2), contradicting the minimality of d. Hence M is irreducible.

Since dM' is defined over R, so is kdM' for any k. On the other hand, suppose mM' is defined over R and write $m = kd + e$, $0 \leq e < d$. We want to show that $e = 0$. Suppose not. Now mM' is defined over R, and so is kdM', and of course $mM' \cong kdM' \oplus eM'$. Then a direct application of Corollary 4.5 (2) shows eM' is defined over R, contradicting the minimality of d. Hence $e = 0$. $\qquad\square$

For future use we record the following:

Proposition 4.18. *Let $\mathbf{F}$ be a finite field, let $\overline{e}_\alpha$ be a central-primitive idempotent of $\overline{R}$, and let $\overline{M}_\alpha$ be the corresponding simple $\overline{R}$-module. If $\overline{e}_\alpha \in R' = \mathbf{E} \otimes_{\mathbf{F}} R \subseteq \overline{R} = \overline{\mathbf{F}} \otimes_{\mathbf{F}} R$, then $\overline{M}_\alpha$ is defined over $\mathbf{E}$.*

Proof. We may assume that $\mathbf{E}$ is a finite extension of $\mathbf{F}$ and hence is finite.

We have that $\overline{R}\overline{e}_\alpha$ is a block $\overline{B}_\alpha$ and so is isomorphic to $n_\alpha \overline{M}_\alpha$ for some integer n_α. Set $B'_\alpha = R'\overline{e}_\alpha$. Then

$$\mathrm{End}_{\overline{R}}(\overline{B}_\alpha) \cong \overline{\mathbf{F}} \otimes_{\mathbf{E}} \mathrm{End}_{R'}(B'_\alpha)$$

so

$$\begin{aligned}
\mathrm{Center}(\mathrm{End}_{\overline{R}}(\overline{B}_\alpha)) &= \mathrm{Center}(\overline{\mathbf{F}} \otimes_{\mathbf{E}} \mathrm{End}_{R'}(B'_\alpha)) \\
&= \overline{\mathbf{F}} \otimes_{\mathbf{E}} \mathrm{Center}(\mathrm{End}_{R'}(B'_\alpha)).
\end{aligned}$$

But $\mathrm{Center}(\mathrm{End}_{\overline{R}}(\overline{B}_\alpha)) = \mathrm{Center}(\mathrm{End}_{\overline{R}}(n_\alpha \overline{M}_\alpha)) \cong \mathrm{End}_{\overline{R}}(\overline{M}_\alpha) = \overline{\mathbf{F}}$ as $\overline{M}_\alpha$ is irreducible and $\overline{\mathbf{F}}$ is algebraically closed. Hence

$$\mathrm{Center}(\mathrm{End}_{R'}(B'_\alpha)) \cong \mathbf{E}.$$

Write $B'_\alpha \cong \bigoplus m_\alpha M'_\alpha$, a simple factorization. Then we see immediately that only one M'_α can appear, and then that $\mathbf{E} \cong \mathrm{Center}(\mathrm{End}_{R'}(B'_\alpha)) \cong \mathrm{Center}(\mathrm{End}_{R'}(M'_\alpha))$.

Now $\mathrm{End}_{R'}(M'_\alpha)$ is a finite division algebra, hence by a famous theorem of Wedderburn, a field (Theorem A.1.1). In particular, it is equal to its center. Thus we see that

$$\mathbf{E} \cong \mathrm{End}_{R'}(M'_\alpha)$$

and hence that M'_α is absolutely irreducible. This implies that $\overline{M}_\alpha = \overline{\mathbf{F}} \otimes_{\mathbf{E}} M'_\alpha$, i.e., that $\overline{M}_\alpha$ is defined over $\mathbf{E}$. $\qquad\square$

Corollary 4.19. *Let $\mathbf{F}$ be finite and $\mathbf{F} \subseteq \mathbf{E} \subseteq \overline{\mathbf{F}}$ where $\mathbf{E}$ is a field such that every central-primitive idempotent of $\overline{R}$ is contained in R'. Then $\mathbf{E}$ is a splitting field for R.*

Proof. By the preceding proposition, $\mathbf{E}$ is a universal field of definition for R. $\qquad\square$

Now that we know that "all the action" occurs between $\mathbf{F}$ and $\overline{\mathbf{F}}$, it becomes interesting to bring Galois theory into play.

Definition 4.20. Let $\mathbf{E} \supseteq \mathbf{F}$ and let M' be an R'-module defined by $\sigma : R' \to \mathrm{Mat}_m(\mathbf{E})$, $m = \dim_{\mathbf{E}} M'$. Let $\gamma \in \mathrm{Gal}(\mathbf{E}/\mathbf{F})$ and define the R'-module $(M')^\gamma$ by $\sigma^\gamma(r) = \sigma(r)^\gamma$. Then M' and $(M')^\gamma$ are **conjugate** R'-modules.

Note that we have chosen a basis in order to define σ, so the definition of $(M')^\gamma$ depends on this choice, but the isomorphism class of $(M')^\gamma$ is well defined. Clearly $(M')^\gamma$ is irreducible if and only if M' is.

Remark 4.21.

(1) We see that $\mathrm{Gal}(\overline{\mathbf{F}}/\mathbf{F})$ acts on the set of irreducible (and hence absolutely irreducible) $\overline{R}$-modules. Indeed, if $\mathbf{E}$ is any splitting field for R, then $\mathrm{Gal}(\mathbf{E}/\mathbf{F})$ acts on the set of irreducible R'-modules. Clearly for any field $\mathbf{E} \subseteq \overline{\mathbf{F}}$, $\mathrm{Gal}(\overline{\mathbf{F}}/\mathbf{E})$ acts trivially on $\overline{R}$-modules that are defined over $\mathbf{E}$. Thus, choosing $\mathbf{E}$ to be a splitting field that is a Galois extension of $\mathbf{F}$, we have that the action of $\mathrm{Gal}(\overline{\mathbf{F}}/\mathbf{F})$ on (absolutely) irreducible $\overline{R}$-modules factors through the action of $\mathrm{Gal}(\overline{\mathbf{F}}/\mathbf{F})/\mathrm{Gal}(\mathbf{E}/\mathbf{F}) = \mathrm{Gal}(\mathbf{E}/\mathbf{F})$ on (absolutely) irreducible R'-modules.

(2) Let $\overline{M}$ be a representation of $\overline{R}$ that is defined over $\mathbf{E}$. Then certainly $(\overline{M})^\gamma = \overline{M}$ for all $\gamma \in \mathrm{Gal}(\overline{\mathbf{F}}/\mathbf{E})$. The converse is also true (consider matrix entries). However, it is *not* in general true that $(\overline{M})^\gamma \cong \overline{M}$ for all $\gamma \in \mathrm{Gal}(\overline{\mathbf{F}}/\mathbf{E})$ implies that M is defined over $\mathbf{E}$. (Thus, we must be careful to distinguish between equality and isomorphism here.)

Proposition 4.22. *Let $\mathbf{E}$ be a finite Galois extension of $\mathbf{F}$ of degree d.*

(1) *Let M' be an R'-module. Then $\bigoplus_{\gamma \in \mathrm{Gal}(\mathbf{E}/\mathbf{F})} (M')^\gamma$ is defined over $\mathbf{F}$.*

(2) *Let M be an irreducible R-module. Then there are integers e and f with $ef = d$, and an irreducible R'-module M', such that M' has exactly f distinct conjugates $\{M'_\beta\}_{\beta \in B}$ and such that $\mathbf{E}(M) \cong e(\bigoplus_{\beta \in B} M'_\beta)$.*

(3) *Let M be an R-module. Then there is an R'-module M' such that $\mathbf{E}(M) \cong \bigoplus_{\gamma \in \mathrm{Gal}(\mathbf{E}/\mathbf{F})} (M')^\gamma$.*

Proof. (1) Choose a normal basis for $\mathbf{E}$ over $\mathbf{F}$ (i.e., a basis of the form $\{\gamma(y) \mid \gamma \in \mathrm{Gal}(\mathbf{E}/\mathbf{F})\}$ for some element y of $\mathbf{E}$).

Let $x \in \mathbf{E}$ such that $\{1, x, \dots, x^{d-1}\}$ is a basis for $\mathbf{E}$ over $\mathbf{F}$, and write $y = \sum_{i=0}^{d-1} f_i x^i$, $f_i \in \mathbf{F}$. Form a d-by-d matrix $P = (p_{ij})$ whose columns are indexed by $\{\gamma \in \mathrm{Gal}(\mathbf{E}/\mathbf{F})\}$ in some arbitrary but fixed order with the first column indexed by $1 \in \mathrm{Gal}(\mathbf{E}/\mathbf{F})$. Let the first column of P have entries $f_0 x^0, f_1 x^1, \dots, f_{d-1} x^{d-1}$, the terms in the above expression for y, and let the column corresponding to γ be constructed similarly from $\gamma(y)$. Thus the action of $\gamma \in \mathrm{Gal}(\mathbf{E}/\mathbf{F})$ on this matrix is by permutation of columns.

Also, P is invertible as $\{\gamma(y)\}$ is a basis for $\mathbf{E}$ over $\mathbf{F}$. Let Q be the md-by-md matrix obtained from P by "inflating" each entry to an m-by-m block, i.e., by replacing p_{ij} by $p_{ij}I$, where I is the m-by-m identity matrix, $m = \dim_{\mathbf{E}} M'$.

Next fix a basis for M' over $\mathbf{E}$. If M' is given by $\sigma : R' \to \mathrm{Aut}(M')$, then, once we have chosen a basis, $\sigma(r)$ is given by a matrix for each $r \in R'$. Fix $r \in R'$ and let $\sigma(r)$ be given by the m-by-m matrix $U = U(r)$. Let $V = V(r)$ be the md-by-md block-diagonal matrix with diagonal blocks U^{γ}, $\gamma \in \mathrm{Gal}(\mathbf{E}/\mathbf{F})$.

Now consider a fixed $\gamma \in \mathrm{Gal}(\mathbf{E}/\mathbf{F})$. As we have observed, the action of γ on the matrix Q is by permuting block columns, so it is given by $Q \mapsto QE$ for some block permutation matrix E. The action of γ on V is given by permuting the diagonal blocks, and this is the action $V \mapsto E^{-1}VE$. Finally, the action of γ on Q^{-1} is given by $Q^{-1} \mapsto (QE)^{-1} = E^{-1}Q^{-1}$. Thus the action of γ on the matrix QVQ^{-1} is given by

$$QVQ^{-1} \mapsto (QE)(E^{-1}VE)(E^{-1}Q^{-1}) = QVQ^{-1}.$$

Thus the matrix QVQ^{-1} is fixed by every $\gamma \in \mathrm{Gal}(\mathbf{E}/\mathbf{F})$, so this matrix has all of its entries in $\mathbf{F}$, and this is true for every $r \in R'$. In other words, the R'-module given by the matrices $\{r \mapsto V(r) \mid r \in R'\}$ is defined over $\mathbf{F}$. But by construction, these matrices are those that define the R'-module $\bigoplus_{\gamma \in \mathrm{Gal}(\mathbf{E}/\mathbf{F})}(M')^{\gamma}$, so this representation is defined over $\mathbf{F}$.

(2) Write $\mathbf{E}(M) = \bigoplus_{\alpha \in A'} n_{\alpha} M'_{\alpha}$ where $A' = \{\alpha \in A \mid n_{\alpha} \neq 0\}$. Then $\mathrm{Gal}(\mathbf{E}/\mathbf{F})$ operates on $\{M'_{\alpha}\}_{\alpha \in A'}$. Let $\{M'_{\beta}\}_{\beta \in B}$ be one orbit. We prove this part of the proposition in two steps. First we show that $\{n_{\beta}\}_{\beta \in B}$ are all equal, and then we show that we have only one orbit, so $B = A'$.

The first step is easy. As $\mathbf{E}(M) = \bigoplus_{\alpha \in A'} n_{\alpha} M'_{\alpha}$, we see that $\mathbf{E}(M)^{\gamma} = \bigoplus_{\alpha \in A'} n_{\alpha} (M'_{\alpha})^{\gamma}$, so the action of $\mathrm{Gal}(\mathbf{E}/\mathbf{F})$ transitively permutes the multiplicities within each orbit. But $\mathbf{E}(M)$ and $\mathbf{E}(M)^{\gamma}$ are isomorphic, as $\mathbf{E}(M)$ is defined over $\mathbf{F}$, so all the multiplicities within each orbit are the same.

The second step requires more work. Consider a single orbit B and let β_1 be any element of B. Then $\bigoplus_{\gamma \in \mathrm{Gal}(\mathbf{E}/\mathbf{F})}(M'_{\beta_1})^{\gamma} \cong \bigoplus_{\beta \in B} n_{\beta} M'_{\beta}$ for some integers $\{n_{\beta}\}$. We have just shown that all of the $\{n_{\beta}\}$ have a common value; call it e, so $\bigoplus_{\beta \in B} eM'_{\beta} \cong \bigoplus_{\gamma \in \mathrm{Gal}(\mathbf{E}/\mathbf{F})}(M'_{\beta_1})^{\gamma}$. By part (1), this implies that $\bigoplus_{\beta \in B} eM'_{\beta} = e(\bigoplus_{\beta \in B} M'_{\beta})$ is defined over $\mathbf{F}$, and moreover, by Proposition 4.17, that for some e_1 dividing e, $e_1(\bigoplus_{\beta \in B} M'_{\beta}) = M'_1 = \mathbf{E}(M_1)$ for some irreducible R-module M_1 defined over $\mathbf{F}$. Now $\mathrm{Hom}_{R'}(\mathbf{E}(M), \mathbf{E}(M_1)) \neq \{0\}$ as both $\mathbf{E}(M)$ and $\mathbf{E}(M_1)$ contain M'_{β_1}. This implies that $\mathrm{Hom}_R(M, M_1) \neq \{0\}$. But M and M_1 are both irreducible, so $M \cong M_1$.

Now suppose that we had at least two orbits, say B_1 and B_2. Then starting from B_1 we construct M_1, as above, and starting from B_2, we construct M_2. Note $\mathrm{Hom}_{R'}(M_1', M_2') = \{0\}$ as they contain no common R'-submodule (since these are distinct orbits), so $\mathrm{Hom}_R(M_1, M_2) = \{0\}$. But this contradicts $M_1 \cong M \cong M_2$.

(3) Note that (2) is a more precise version of (3) in the case that M is irreducible, so, by taking direct sums, (2) immediately implies (3). $\quad\square$

Corollary 4.23. *Let $\mathbf{E}$ be a Galois extension of $\mathbf{F}$ and let M' be an irreducible R'-module such that $(M')^\gamma$ is isomorphic to M' for every $\gamma \in \mathrm{Gal}(\mathbf{E}/\mathbf{F})$. Then dM' is defined over $\mathbf{F}$ for some d dividing $[\mathbf{E} : \mathbf{F}]$.*

Proof. Immediate from Propositions 4.17 and 4.22. $\quad\square$

We are concerned here with representations of algebras over fields. However, it is natural to ask the question of integrality.

Before answering this question we need to introduce a bit of background material. An integral domain I with the property that every finitely generated I-ideal is principal is called a **Bézout domain**. The basic theorem that every finitely generated torsion-free module over a PID is free generalizes unchanged to modules over Bézout domains: every finitely generated torsion-free module over a Bézout domain is free (Theorem A.1.3).

Proposition 4.24. *Let I be a Bézout domain with quotient field $\mathbf{F}$. Let R_0 be an I-algebra that is of finite rank, and let $R = \mathbf{F} \otimes_I R_0$. Then every R-module M with $\dim_{\mathbf{F}} M$ finite is isomorphic to $\mathbf{F} \otimes_I N$ for some R_0-module N that is free as an I-module.*

Proof. (Before beginning the proof, observe that M is certainly free as an $\mathbf{F}$-module as $\mathbf{F}$ is a field.) Choose a basis $\{i_1, \dots, i_d\}$, $d = \mathrm{rank}_I R_0$, for R_0 over I, and a basis $\{m_1, \dots, m_e\}$, $e = \dim_{\mathbf{F}} M$, for M over $\mathbf{F}$. Let N be the I-module generated by $\{i_s m_t\}$. Then N is visibly finitely generated. Also, M is certainly $\mathbf{F}$-torsion-free, as $\mathbf{F}$ is a field, and $N \subseteq M$, so N is I-torsion-free. Hence N is free as an I-module, and clearly $M = \mathbf{F} \otimes_I N$. $\square$

Corollary 4.25. *In the situation of the preceding proposition, let the R-module structure on M be given by $\sigma : R \to \mathrm{Aut}_{\mathbf{F}}(M)$.*

(1) *If $\mathbf{F} = \mathbf{Q}$, then M has a basis in which all of the entries of the matrices $[\sigma(i_s)]$ are integers.*

(2) *If $\mathbf{F} = \overline{\mathbf{Q}}$, then M has a basis in which all of the entries of the matrices $[\sigma(i_s)]$ are algebraic integers.*

(3) *If I is a discrete valuation ring with quotient field $\mathbf{F}$, then M has a basis in which all of the entries of the matrices $[\sigma(i_s)]$ are elements of I.*

Proof. Part (1) is immediate from Proposition 4.24 and the fact that $\mathbb{Z}$ is a PID. Part (2) is immediate from Proposition 4.24 and the fact that the ring of all algebraic integers is a Bézout domain, i.e., that every finitely generated ideal in the ring of all algebraic integers is principal (Theorem A.1.2). Part (3) is immediate from Proposition 4.24 and the fact that every discrete valuation ring (DVR) is a PID. (See [Ma, Section 11] for the definition of a DVR and the proof of this fact.) $\square$

2.5. Theorems of Burnside and Frobenius-Schur

In this section we assume that R is a semisimple ring that is a finite-dimensional algebra over $\mathbf{F}$*, and that* $\mathbf{F}$ *is a splitting field for R.*

First observe that we proved Corollary 2.24 under the assumption that $\mathbf{F}$ is algebraically closed. That assumption is stronger than necessary.

Theorem 5.1. *Corollary 2.24 holds under the assumption that* $\mathbf{F}$ *is a splitting field for R.*

Proof. By the definition of a splitting field, every irreducible R-module M is absolutely irreducible so, by Corollary 4.8, $\mathrm{End}_R(M) \cong \mathbf{F}$ for every M, and then this is immediate from Theorem 2.22. $\square$

For an R-module M, we will let $\sigma : R \to \mathrm{End}_\mathbf{F}(M)$ be the map given by $\sigma(r)(m) = rm$.

Definition 5.2. M is a **faithful** R-module if $\mathrm{Ker}(\sigma) = \{0\}$.

Theorem 5.3 (Burnside). *Let M be a simple R-module, and let $\sigma : R \to \mathrm{End}_\mathbf{F}(M)$ be as above. Then σ is a surjection, i.e., $\sigma(R) = \mathrm{End}_\mathbf{F}(M)$.*

Proof. Let $n = \dim_\mathbf{F} R$.

First consider R as a R-module, giving a map $\sigma_0 : R \to \mathrm{End}_\mathbf{F}(R)$ (by $\sigma_0(r)(s) = rs$). Clearly σ_0 is faithful, so $\dim_\mathbf{F} \sigma_0(R) = n$.

We have a simple factorization $R \cong \bigoplus_{\alpha \in A} n_\alpha M_\alpha$ so $\sigma_0 \cong \bigoplus_{\alpha \in A} n_\alpha \sigma_\alpha$. But

$$n = \dim_\mathbf{F} \sigma_0(R) = \sum_\alpha \dim_\mathbf{F} \sigma_\alpha(R) \le \sum_\alpha \dim_\mathbf{F} \mathrm{End}_\mathbf{F}(M_\alpha) = \sum_\alpha n_\alpha^2.$$

However, by Theorem 5.1, $n = \sum_\alpha n_\alpha^2$, so we must have $\dim_\mathbf{F} \sigma_\alpha(R) = \dim_\mathbf{F} \mathrm{End}_\mathbf{F}(M_\alpha)$, and hence $\sigma_\alpha(R) = \mathrm{End}_\mathbf{F}(M_\alpha)$, for each α. $\square$

Corollary 5.4. *For any basis* $\{r_1, \ldots, r_n\}$ *of* R *over* $\mathbf{F}$,

$$\{\sigma(r_1), \ldots, \sigma(r_n)\}$$

spans $\mathrm{End}_{\mathbf{F}}(M)$.

Proof. $\qquad\qquad\qquad\qquad\qquad\qquad\qquad\qquad\qquad\qquad\qquad\qquad\qquad\square$

Lemma 5.5. *Let* $R \cong \bigoplus_{\alpha \in A} n_\alpha M_\alpha$ *and suppose that* $N \cong \bigoplus_{\alpha \in A} n'_\alpha M_\alpha$. *Then* N *is a faithful* R*-module if and only if* $n'_\alpha \geq 1$ *for each* α *in* A.

Proof. Suppose $n_\alpha \geq 1$ for each α. Then for some k, R as an R module is isomorphic to a submodule of N^k. But then R faithful implies N^k faithful, which is equivalent to N faithful.

On the other hand, suppose $n_{\alpha_0} = 0$ for some α_0. Writing $R = \bigoplus_{i \in I} M_i$, a direct sum of simple R-modules, let $S = \bigoplus_{j \in J} M_j$, where the sum for S is taken over those M_j not isomorphic to M_{α_0}. Let $m \in M_i$ for some i with M_i isomorphic to M_{α_0}. By Schur's lemma, $\mathrm{Hom}_R(M_i, S) = \{0\}$, so $\sigma(m)(s) = ms = 0$ for every $s \in S$, and so S is not faithful. But then N is a quotient of S^k for some k, so S not faithful is equivalent to S^k not faithful, which implies N not faithful. $\qquad\qquad\qquad\qquad\qquad\qquad\qquad\square$

Theorem 5.6 (Frobenius-Schur). *Let* $\{M_\alpha\}_{\alpha \in A}$ *be representatives of the isomorphism classes of simple* R*-modules. Pick bases for each* M_α *and consider* $\{f_{ij}^\alpha \mid i, j = 1, \ldots, n_\alpha, \alpha \in A\}$, *where* f_{ij}^α *are the matrix entries of* σ_α. *Then* $\{f_{ij}^\alpha\}$ *is a linearly independent set of functions on* R.

Proof. Let $n = \dim_{\mathbf{F}} R$.

Set $M = \bigoplus_{\alpha \in A} M_\alpha$, and let M be given by $\sigma : R \to \mathrm{End}_{\mathbf{F}}(M)$. Note $\sigma = \bigoplus_{\alpha \in A} \sigma_\alpha$. Then $\sigma(R) \subseteq \bigoplus_{\alpha \in A} \sigma_\alpha(R) \subseteq \bigoplus_{\alpha \in A} \mathrm{End}_{\mathbf{F}}(M_\alpha)$. However, by Lemma 5.5, σ is faithful, so

$$n = \dim_{\mathbf{F}} R = \dim_{\mathbf{F}} \sigma(R) \leq \sum_{\alpha \in A} \dim_{\mathbf{F}}(\mathrm{End}_{\mathbf{F}}(M_\alpha))$$

$$= \sum_{\alpha \in A} (\dim_{\mathbf{F}} M_\alpha)^2 = \sum_{\alpha \in A} n_\alpha^2.$$

But $n = \sum_{\alpha \in A} n_\alpha^2$ so we must have equality, and $\sigma(R) = \bigoplus_{\alpha \in A} \mathrm{End}_{\mathbf{F}}(M_\alpha)$. From this the theorem follows readily: Suppose $\sum c_{ij}^\alpha f_{ij}^\alpha(r) = 0$ for all $r \in R$. Pick $r \in R$ with $\sigma(r)$ having only a single nonzero entry—the (i, j) entry in σ_α. Then $c_{ij}^\alpha = 0$. $\qquad\qquad\qquad\qquad\square$

Note that there are a total of n functions $\{f_{ij}^\alpha\}$. Thus, choosing an arbitrary (but fixed) ordering of them, we may consider $f(r) = \{f_{ij}^\alpha(r)\}$ to be an element of $\mathbf{F}^n$, for every $r \in R$.

Corollary 5.7. *For any basis $\{r_1, \ldots, r_n\}$ of R over $\mathbf{F}$, the n vectors*

$$\{f(r_1), \ldots, f(r_n)\}$$

are linearly independent.

Proof. $\qquad\qquad\qquad\qquad\qquad\qquad\qquad\qquad\qquad\qquad\qquad\qquad\qquad\quad\square$

Chapter 3

Semisimple Group Representations

3.1. Examples and General Results

For the sake of continuity and ease of exposition, we begin by recapitulating some of the definitions and examples in the introduction.

Definition 1.1. Let G be a group and $\mathbf{F}$ a field. An **F-representation** of G is a (left) $\mathbf{F}(G)$-module M.

In other words, M is an $\mathbf{F}$-vector space, and for each $g \in G$ we have a linear transformation $\sigma(g) : M \to M$ given by the action of g regarded as $1g \in \mathbf{F}(G)$ on M. These linear transformations satisfy $\sigma(1) = 1_M$ and $\sigma(g_2 g_1)(m) = \sigma(g_2)(\sigma(g_1)(m))$ for all $m \in M$ and $g \in G$. Note then that $\sigma(g^{-1})\sigma(g) = \sigma(1) = 1_M$ so that $\sigma(g^{-1}) = \sigma(g)^{-1}$. In particular, each $\sigma(g)$ is invertible, i.e., $\sigma(g) \in \mathrm{Aut}(M)$.

Conversely, we may view an $\mathbf{F}$-representation of G on M, where M is an $\mathbf{F}$-vector space, as being given by a homomorphism $\sigma : G \to \mathrm{Aut}(M)$. Then we define an $\mathbf{F}(G)$-module structure on M by

$$\left(\sum_{g \in G} a_g g \right) (m) = \sum_{g \in G} a_g \sigma(g)(m).$$

In this situation, we say that the representation is defined by σ.

We will denote a representation as above by M when we wish to emphasize the underlying vector space or (more often) by σ when we wish to emphasize the homomorphism, and we will use the term representation instead of **F**-representation when **F** is understood. Occasionally (as is often done) we shall omit σ when it is understood and write $g(m)$ for $\sigma(g)(m)$.

Definition 1.2. The **degree** $\deg(M)$ of a representation M is

$$\dim_{\mathbf{F}}(M) \in \{0, 1, 2, \dots\} \cup \{\infty\}.$$

Two **F**-representations M_1 and M_2 of G, defined by $\sigma_i : G \to \mathrm{Aut}(M_i)$ for $i = 1, 2$, are said to be **isomorphic (or equivalent)** if they are isomorphic as **F**(G)-modules. Concretely, this is the case if and only if there is an invertible **F**-linear transformation $f : M_1 \to M_2$ with

$$f(\sigma_1(g)(m)) = \sigma_2(g)(f(m)) \text{ for every } m \in M, \ g \in G.$$

We will be considering general groups G and fields **F**, though we will obtain our strongest results when G is finite (and **F** satisfies certain restrictions). *Accordingly, and with a bit of hindsight, we will adopt the following notational convention henceforth:*

Notation. Let G be a finite group. Then

(1) n denotes the order of G,

(2) t denotes the number of conjugacy classes of elements of G,

(3) u denotes the exponent of G.

Since we are considering **F**(G)-modules, we make some elementary remarks about **F**(G) itself.

(1) **F**(G) is an **F**-algebra.

(2) **F**(G) is an (**F**(G), **F**(G))-bimodule, as well as an (**F**(K), **F**(H))-bimodule for any pair of subgroups K and H of G.

(3) **F**(G) is commutative if and only if G is abelian.

(4) If G has torsion (i.e., elements of finite order), then **F**(G) has zero divisors, as if $g \in G$ with $g^m = 1$, then

$$(1 - g)(1 + g + \cdots + g^{m-1}) = 0 \in \mathbf{F}(G).$$

(5) Any **F**(G)-module is an (**F**(G), **F**)-bimodule.

It will be useful to us to single out the following class of fields **F**.

Definition 1.3. Let G be a finite group. A field **F** is called **good for** G (or simply **good**) if the following conditions are satisfied.

(1) The characteristic of $\mathbf{F}$ is 0 or relatively prime to the order of G.

(2) If u denotes the exponent of G, then the equation $X^u - 1 = 0$ has u *distinct* roots in $\mathbf{F}$.

Remark. Actually, (2) implies (1), and also (2) implies that the equation $X^k - 1 = 0$ has k distinct roots in $\mathbf{F}$, for every k dividing u. Furthermore, since the roots of $X^k - 1 = 0$ form a subgroup of the multiplicative group $\mathbf{F}^*$, which is cyclic, there is (at least) one root $\zeta = \zeta_k$ such that these roots are

$$1 = \zeta^0, \zeta, \zeta^2, \ldots, \zeta^{k-1}.$$

We shall reserve the use of the symbol ζ (or ζ_k) to denote this. We further assume that these roots have been consistently chosen, in the following sense:

$$\text{If } k_1 \text{ divides } k_2, \text{ then } (\zeta_{k_2})^{k_2/k_1} = \zeta_{k_1}.$$

Note that the field $\mathbf{C}$ (or, in fact, any algebraically closed field of characteristic zero) is good for every finite group, and in $\mathbf{C}$ we may simply choose $\zeta_k = \exp(2\pi i/k)$ for every positive integer k.

In this chapter we will have occasion to consider Hom_R, End_R, or $\otimes_R$ for various rings of the form $R = \mathbf{F}(H)$, where H is a subgroup of G (or $R = \mathbf{F}$, in which case we may identify R with $\mathbf{F}(\{1\})$). We adopt the notational convention that Hom, End, and $\otimes$ mean $\mathrm{Hom}_{\mathbf{F}}$, $\mathrm{End}_{\mathbf{F}}$, and $\otimes_{\mathbf{F}}$, respectively, and Hom_H, End_H, and $\otimes_H$ mean Hom_R, End_R, and $\otimes_R$, respectively, for $R = \mathbf{F}(H)$.

Example 1.4.

(1) Let M be an $\mathbf{F}$-vector space of dimension 1 and define $\sigma : G \to \mathrm{Aut}(M)$ by $\sigma(g) = 1_M$ for all $g \in G$. We call M the **trivial representation of degree** 1, and we denote it by τ.

(2) $M = \mathbf{F}(G)$ as an $\mathbf{F}(G)$-module. This is a representation of degree n. As an $\mathbf{F}$-vector space, M has a basis $\{g : g \in G\}$, and an element $g_0 \in G$ acts on M by

$$\sigma(g_0)\left(\sum_{g\in G} a_g g\right) = \sum_{g\in G} a_g g_0 g.$$

M is called the **(left) regular representation** of G and it plays a crucial role in the theory. We denote it by $\mathcal{R} = \mathcal{R}(G)$.

(3) **Permutation representations.** Let P be a set, $P = \{p_i\}_{i\in I}$, and suppose we have a homomorphism $\sigma : G \to S_P = \mathrm{Aut}(P)$. Let $M = \mathbf{F}(P)$ be the free $\mathbf{F}$-module with basis P. Then G acts on $\mathbf{F}(P)$ by the formula

$$\sigma(g)\left(\sum_{i\in I} a_i p_i\right) = \sum_{i\in I} a_i(\sigma(g)(p_i)),$$

giving a representation of degree $|P|$. Note that the regular representation is a special case of this construction, obtained by taking $P = G$. As an important variant, we could take $P = G/H = \{gH\}$, the set of left cosets of some subgroup H of G.

(4) If M_1 and M_2 are two representations of G, defined by σ_1 and σ_2, then $M_1 \oplus M_2$ is a representation of G defined by $\sigma_1 \oplus \sigma_2$, and $\deg(M_1 \oplus M_2) = \deg(M_1) + \deg(M_2)$.

(5) If M_1 and M_2 are two representations of G, defined by σ_1 and σ_2, then $M_1 \otimes M_2$ is a representation of G defined by $\sigma_1 \otimes \sigma_2$, and $\deg(M_1 \otimes M_2) = (\deg(M_1))(\deg(M_2))$.

(6) Let $G = \mathbb{Z}/n\mathbb{Z} = \langle g \mid g^n = 1 \rangle$ be cyclic of order n and let $\mathbf{F}$ be a field that is good for G. We have the 1-dimensional representations $\theta_k : G \to \mathrm{Aut}(\mathbf{F}) \cong \mathbf{F}^*$ defined by

$$\theta_k(g) = \zeta^k \text{ for } k = 0, \dots, n - 1.$$

These are all distinct (i.e., pairwise nonisomorphic) and $\theta_0 = \tau$.

(7) Let

$$G = D_{2m} = \langle x, y \mid x^m = 1, y^2 = 1, xy = yx^{-1} \rangle$$

be the dihedral group of order $2m$, and let $\mathbf{F}$ be a field that is good for G. The representations of D_{2m} are described in tables 1.1 and 1.2, using the following matrices:

$$(1.1) \qquad A_k = \begin{bmatrix} \zeta^k & 0 \\ 0 & \zeta^{-k} \end{bmatrix} \quad \text{and} \quad B = \begin{bmatrix} 0 & 1 \\ 1 & 0 \end{bmatrix}.$$

(Note that it is only necessary to give the value of the representation on the two generators x and y.)

Table 1.1. Representations of D_{2m} (m odd)

Representation	x	y	degree	
$\psi_+ = \tau$	1	1	1	
ψ_-	1	-1	1	
φ_k	A_k	B	2	$1 \leq k \leq (m-1)/2$

Table 1.2. Representations of D_{2m} (m even)

Representation	x	y	degree	
$\psi_{++} = \tau$	1	1	1	
ψ_{+-}	1	-1	1	
ψ_{-+}	-1	1	1	
ψ_{--}	-1	-1	1	
φ_k	A_k	B	2	$1 \leq k \leq (m/2) - 1$

(8) Suppose that M has a basis $\mathcal{B}$ such that for every $g \in G$, $[\sigma(g)]_{\mathcal{B}}$ is a matrix with exactly one nonzero entry in every row and column. Then M is called a **monomial representation of** G. For example, the representations of D_{2m} given above are monomial. If all nonzero entries of $[\sigma(g)]_{\mathcal{B}}$ are 1, then the monomial representation is called a **permutation representation**. (The reader should check that this definition agrees with the definition of permutation representation given in part (3) of this example.)

(9) Let $X = \{1, x, x^2, \dots\}$ with the multiplication $x^i x^j = x^{i+j}$. Then X is a **monoid**, i.e., it satisfies all of the group axioms except for the existence of inverses. Then one can define the monoid ring exactly as in the case of the group ring. If this is done, then $\mathbf{F}(X)$ is just the polynomial ring $\mathbf{F}[x]$. Let M be an $\mathbf{F}$ vector space and let $T : M \to M$ be a linear transformation. Then M becomes an $\mathbf{F}(X)$-module via $x^i(m) = T^i(m)$, for $m \in M$ and $i \geq 0$. Thus, we have an example of a monoid representation. Now we may identify $\mathbb{Z}$ with

$$\{\dots, x^{-2}, x^{-1}, 1, x, x^2, \dots\}.$$

If the linear transformation T is invertible, then M becomes an $\mathbf{F}(\mathbb{Z})$-module via the action $x^i(m) = T^i(m)$ for $m \in M$, $i \in \mathbb{Z}$.

(10) As an example of (9), let $T : \mathbf{F}^2 \to \mathbf{F}^2$ have matrix $\left[\begin{smallmatrix} 1 & 1 \\ 0 & 1 \end{smallmatrix}\right]$ (in the standard basis). Then we obtain a representation of $\mathbb{Z}$ of degree 2. If $\operatorname{char}(\mathbb{Z}) = p > 0$, then $T^p = 1_{\mathbf{F}^2}$, so in this case we obtain a representation of $\mathbb{Z}/p\mathbb{Z}$ of degree 2.

(11) Let $\varepsilon : \mathcal{R}(G) \to \mathbf{F}$ be defined by

$$\varepsilon\left(\sum_{g \in G} a_g g\right) = \sum_{g \in G} a_g.$$

If we let G act trivially on $\mathbf{F}$, we may regard ε as a map of $\mathbf{F}$ representations

$$\varepsilon : \mathcal{R}(G) \to \tau.$$

We let $\mathcal{R}_0(G) = \operatorname{Ker}(\varepsilon)$. The homomorphism ε is known as the **augmentation map** and $\mathcal{R}_0(G)$ is known as the **augmentation ideal** of $\mathcal{R}(G)$. It is then, of course, an $\mathbf{F}$-representation of G.

(12) If $\mathbf{F}$ is a subfield of $\mathbf{E}$ and M is an $\mathbf{F}$-representation of G, then $M' = \mathbf{E} \otimes_{\mathbf{F}} M$ is an $\mathbf{E}$-representation of G. An $\mathbf{E}$-representation arising in this way is said to be **defined over** $\mathbf{F}$.

(13) If M_i is an $\mathbf{F}$-representation of G_i, defined by σ_i for $i = 1, 2$, then $M_1 \oplus M_2$ is a representation of $G_1 \times G_2$, defined by

$$(\sigma_1 \oplus \sigma_2)(g_1, g_2) = \sigma_1(g_1) \oplus \sigma_2(g_2) \quad \text{for } g_i \in G_i.$$

(Compare with part (4) of this example.)

(14) If M_i is an **F**-representation of G_i, defined by σ_i for $i = 1, 2$, then $M_1 \otimes M_2$ is a representation of $G_1 \times G_2$, defined by

$$(\sigma_1 \otimes \sigma_2)(g_1, g_2) = \sigma_1(g_1) \otimes \sigma_2(g_2) \quad \text{for } g_i \in G_i.$$

(Compare with part (5) of this example.)

(15) If $\sigma : G \to \mathrm{Aut}(M)$ is injective, then σ is called **faithful**. If not, let $K = \mathrm{Ker}(\sigma)$. Then σ determines a homomorphism $\sigma' : G/K \to \mathrm{Aut}(M)$, which is a faithful representation of the group G/K.

(16) Let $f : G_1 \to G_2$ be a group homomorphism and let $\sigma : G_2 \to \mathrm{Aut}(M)$ be a representation of G_2. The **pullback of σ by f**, denoted $f^*(\sigma)$, is the representation of G_1 defined by $f^*(\sigma) = \sigma \circ f$, i.e., $f^*(\sigma) : G_1 \to \mathrm{Aut}(M)$ by

$$f^*(\sigma)(g) = \sigma(f(g)) \quad \text{for } g \in G_1.$$

(17) Let M be an **F**-representation of G, so M is an $\mathbf{F}(G)$-module. Then for any subgroup H of G, M is also an $\mathbf{F}(H)$-module, i.e., an **F**-representation of H.

(18) Let H be a subgroup of G and let M be an **F**-representation of H, i.e., an $\mathbf{F}(H)$-module. Then $\mathbf{F}(G) \otimes_{\mathbf{F}(H)} M$ is an $\mathbf{F}(G)$-module, i.e., an **F**-representation of G. This **F**-representation of G is said to be **induced** from the given **F**-representation of H.

(19) Let M be an **F**-representation of G, defined by σ, and let γ be an automorphism of $\mathbf{F}(G)$. Then M^γ is the **F**-representation of G defined by σ^γ, where $\sigma^\gamma(r)(m) = \sigma(\gamma(r))(m)$, $r \in \mathbf{F}(G)$ and $m \in M$. Note in particular that γ may arise from either an automorphism of the field **F** or an automorphism of the group G. If γ arises from an automorphism of **F**, M and M^γ are called **conjugate** representations.

(We have seen many of these examples before, and we will see all of them later. We leave the verification of these examples to the reader.)

Remark. There are a couple of small differences between common terminology in module theory and representation theory. To say an R-module M is trivial means $M = \{0\}$; to say a G-representation is trivial means only that $\sigma(g)(m) = m$ for all g in G and m in M. To say an R-module M is faithful means that $rm = m$ for all $m \in M$ implies that $r = 1$; to say that a G-module M is faithful means that $\sigma(g)(m) = m$ for all $m \in M$ implies that $g = 1$, and this is weaker.

Definition 1.5.

(1) M is an **irreducible representation** of G if M is an irreducible $\mathbf{F}(G)$-module.

(2) M is an **indecomposable representation** of G if M is an indecomposable $\mathbf{F}(G)$-module.

(3) M is a **semisimple representation** of G if M is a semisimple $\mathbf{F}(G)$-module.

One of our principal objectives will be to find irreducible representations of a group G and to show how to express a representation as a sum of irreducible representations, when possible. The following examples illustrate this theme.

Example 1.6. *Let $\mathbf{F}$ be good for $\mathbb{Z}/n\mathbb{Z}$. Then the regular representation $\mathcal{R}(\mathbb{Z}/n\mathbb{Z})$ is isomorphic to*

$$\theta_0 \oplus \theta_1 \oplus \cdots \oplus \theta_{n-1}.$$

Proof. Consider the $\mathbf{F}$-basis $\{1, g, \dots, g^{n-1}\}$ of $\mathcal{R}(\mathbb{Z}/n\mathbb{Z})$. In this basis, $\sigma(g)$ has the matrix

$$\begin{bmatrix} 0 & 0 & \cdots & 0 & 1 \\ 1 & 0 & \cdots & 0 & 0 \\ 0 & 1 & \cdots & 0 & 0 \\ \vdots & \vdots & \ddots & \vdots & \vdots \\ 0 & 0 & \cdots & 0 & 0 \\ 0 & 0 & \cdots & 1 & 0 \end{bmatrix},$$

which we recognize as $C(X^n - 1)$, the companion matrix of the polynomial $X^n - 1$. Hence both the minimum and characteristic polynomials of $\sigma(g)$ are $X^n - 1$. By our assumption on $\mathbf{F}$, this polynomial factors into distinct linear factors

$$X^n - 1 = \prod_{k=0}^{n-1} (X - \zeta^k).$$

Hence, $\sigma(g)$ is diagonalizable, and indeed, in a proper basis $\mathcal{B}$, it has matrix

$$\operatorname{diag}(1, \zeta, \zeta^2, \dots, \zeta^{n-1}).$$

Each of the eigenspaces is an $\mathbf{F}(\mathbb{Z}/n\mathbb{Z})$-submodule, so we see immediately that

$$\mathcal{R}(\mathbb{Z}/n\mathbb{Z}) \cong \theta_0 \oplus \theta_1 \oplus \cdots \oplus \theta_{n-1}.$$

$\square$

Example 1.7. Let $\mathbf{F} = \mathbf{Q}$, the rational numbers, and let p be a prime. Consider the augmentation ideal $\mathcal{R}_0 \subseteq \mathcal{R} = \mathbf{Q}(\mathbb{Z}/p\mathbb{Z})$. If g is a generator of $\mathbb{Z}/p\mathbb{Z}$ and $T = \sigma(g) : \mathcal{R}_0 \to \mathcal{R}_0$, then its minimum polynomial $m_T(X)$ and its characteristic polynomial $c_T(X)$ are given by

$$m_T(X) = c_T(X) = \frac{(X^p - 1)}{(X - 1)},$$

which is irreducible. Hence, $\mathcal{R}_0$ is an irreducible $\mathbf{Q}$-representation of $\mathbb{Z}/p\mathbb{Z}$. Note that

$$\mathbf{C} \otimes_{\mathbf{Q}} \mathcal{R}_0$$

is the augmentation ideal in $\mathbf{C}(G)$, and $\mathbf{C} \otimes_{\mathbf{Q}} \mathcal{R}_0$ is not irreducible and, in fact, is isomorphic to

$$\theta_1 \oplus \cdots \oplus \theta_{p-1}.$$

(In fact, this statement is true without the requirement that p be prime.)

Example 1.8. Let $\mathbf{F}$ be good for D_{2m}.

(1) Each of the $\mathbf{F}(D_{2m})$-modules φ_k of Example 1.4 (7) is irreducible, and they are distinct.

(2) If, for m even, we define $\varphi_{m/2}$ by $\varphi_{m/2}(x) = A_{m/2}$ and $\varphi_{m/2}(y) = B$, then

$$\varphi_{m/2} \cong \psi_{-+} \oplus \psi_{---}.$$

(3) Let P be the set of vertices of a regular m-gon, and let D_{2m} act on P in the usual manner. Then, as $\mathbf{F}(D_{2m})$-modules,

$$\mathbf{F}(P) \cong \begin{cases} \psi_+ \oplus \varphi_1 \oplus \varphi_2 \oplus \cdots \oplus \varphi_{(m-1)/2} & \text{for } m \text{ odd,} \\ \psi_{++} \oplus \psi_{-+} \oplus \varphi_1 \oplus \varphi_2 \oplus \cdots \oplus \varphi_{\frac{m}{2}-1} & \text{for } m \text{ even.} \end{cases}$$

(4) For the regular representation, we have

$$\mathcal{R}(D_{2m}) \cong \begin{cases} \psi_+ \oplus \psi_- \oplus 2\varphi_1 \oplus \cdots \oplus 2\varphi_{(m-1)/2} & \text{for } m \text{ odd,} \\ \psi_{++} \oplus \psi_{+-} \oplus \psi_{-+} \oplus \psi_{--} \oplus 2\varphi_1 \oplus \cdots \oplus 2\varphi_{\frac{m}{2}-1} & \text{for } m \text{ even.} \end{cases}$$

We leave the verification of these claims as an exercise for the reader.

Definition 1.9. Let M and N be $\mathbf{F}$-representations of G. The **multiplicity m of M in N** is the largest nonnegative integer m with the property that mM is isomorphic to a submodule of N. If no such m exists, i.e., if mM is isomorphic to a submodule of N for every nonnegative integer m, we say that the multiplicity of M in N is infinite.

If the multiplicity of M in N is m, we shall often say that N contains M m times (or that N contains m copies of M). If $m = 0$, we shall often say that N does not contain M.

Lemma 1.10. *Let G be a group. Then the regular representation $\mathcal{R}$ contains the trivial representation τ once if G is finite, but it does not contain τ if G is infinite.*

Proof. Let M be the submodule of $\mathcal{R}$ consisting of all elements on which G acts trivially. Let $m \in M$. Then we may write

$$m = \sum_{g \in G} a_g g \quad \text{where } a_g \in \mathbf{F}.$$

By assumption, $g_0 m = m$ for every $g_0 \in G$. But

$$g_0 m = \sum_{g \in G} a_g (g_0 g)$$

and

$$m = \sum_{g \in G} a_{g_0 g}(g_0 g).$$

Therefore, $a_{g_0 g} = a_g$ for every $g \in G$ and every $g_0 \in G$. In particular,

$$(1.2) \qquad a_{g_0} = a_1$$

for every $g_0 \in G$.

Now suppose that G is finite. Then, by equation (1.2),

$$m = \sum_{g \in G} a_1 g = a_1 \left(\sum_{g \in G} g \right) \quad \text{where } a_1 \in \mathbf{F},$$

i.e., $M = \langle \sum_{g \in G} g \rangle$ as an $\mathbf{F}$-module. Therefore, M is 1-dimensional over $\mathbf{F}$, so that $M = \tau$.

On the other hand, if G is infinite, equation (1.2) implies that $a_g \neq 0$ for all $g \in G$ whenever $a_1 \neq 0$. But the group ring $\mathbf{F}(G)$ consists of *finite* sums $\sum_{g \in G} a_g g$, so this is impossible. Hence, we must have $a_1 = 0$ and $M = \{0\}$. $\square$

Corollary 1.11. *If G is an infinite group, then $\mathcal{R}$ is not semisimple.*

Proof. Suppose that $\mathcal{R}$ were semisimple. Then by Theorem 2.2.14, $\mathcal{R}$ would contain the simple $\mathcal{R}$-module τ, but by Lemma 1.10, it does not. $\square$

Example 1.12. Consider Example 1.4 (10). This is an $\mathbf{F}$-representation of $\mathbb{Z}$, which is indecomposable but not irreducible: it has τ as a subrepresentation (consisting of vectors in $\mathbf{F} \times \{0\}$). Thus this representation is not semisimple.

The following lemma, incorporating the technique of **averaging over the group**, is crucial.

Lemma 1.13. *Let G be a finite group and $\mathbf{F}$ a field with $\mathrm{char}(\mathbf{F}) = 0$ or prime to the order of G. Let V_1 and V_2 be $\mathbf{F}$-representations of G defined by $\sigma_i : G \to \mathrm{Aut}(V_i)$ for $i = 1, 2$, and let $n = |G|$. Let $f \in \mathrm{Hom}(V_1, V_2)$ and set*

$$(1.3) \qquad \mathrm{Av}(f) = \frac{1}{n} \sum_{g \in G} \sigma_2(g^{-1}) f(\sigma_1(g)).$$

Then $\mathrm{Av}(f) \in \mathrm{Hom}_G(V_1, V_2)$. *Furthermore, if* $f \in \mathrm{Hom}_G(V_1, V_2)$, *then*

$$(1.4) \qquad\qquad \mathrm{Av}(f) = f.$$

Proof. We need to show that for every $g_0 \in G$ and every $v_1 \in V_1$,

$$\sigma_2(g_0) \, \mathrm{Av}(f)(v_1) = \mathrm{Av}(f)(\sigma_1(g_0)(v_1)).$$

But

$$
\begin{aligned}
\mathrm{Av}(f)(\sigma_1(g_0)(v_1)) &= \left(\frac{1}{n} \sum_{g \in G} \sigma_2(g^{-1}) f(\sigma_1(g)) \right) \left(\sigma_1(g_0)(v_1) \right) \\
&= \frac{1}{n} \sum_{g \in G} \sigma_2(g^{-1}) f(\sigma_1(g)) \left(\sigma_1(g_0)(v_1) \right) \\
&= \frac{1}{n} \sum_{g \in G} \sigma_2(g^{-1}) f(\sigma_1(g g_0))(v_1) \\
&= \frac{1}{n} \sum_{g \in G} \sigma_2(g_0) \sigma_2(g_0^{-1} g^{-1}) f(\sigma_1(g g_0))(v_1) \\
&= \frac{1}{n} \sum_{g \in G} \sigma_2(g_0) \sigma_2((g g_0)^{-1}) f(\sigma_1(g g_0))(v_1).
\end{aligned}
$$

Let $g' = g g_0$. As g runs through the elements of G, so does g'. Thus,

$$
\begin{aligned}
\mathrm{Av}(f)(\sigma_1(g_0)(v_1)) &= \frac{1}{n} \sum_{g' \in G} \sigma_2(g_0) \sigma_2(g'^{-1}) f(\sigma_1(g'))(v_1) \\
&= \sigma_2(g_0) \left(\frac{1}{n} \sum_{g' \in G} \sigma_2(g'^{-1}) f(\sigma_1(g')) \right) (v_1) \\
&= \sigma_2(g_0) \, \mathrm{Av}(f)(v_1)
\end{aligned}
$$

as required, so equation (1.3) is satisfied.

Also, if $f \in \mathrm{Hom}_G(V_1, V_2)$, then for every $g \in G$,

$$\sigma_2(g) f = f \sigma_1(g).$$

Hence, in this case

$$\mathrm{Av}(f) = \frac{1}{n} \sum_{g \in G} \sigma_2(g^{-1}) f \sigma_1(g)$$

$$= \frac{1}{n} \sum_{g \in G} f \sigma_1(g^{-1}) \sigma_1(g)$$

$$= \frac{1}{n} \sum_{g \in G} f \sigma_1(e)$$

$$= \frac{1}{n} \sum_{g \in G} f$$

$$= \frac{1}{n}(nf)$$

$$= f.$$

Hence, equation (1.4) is satisfied. $\qquad\square$

Now we come to one of the cornerstones of the representation theory of finite groups.

Theorem 1.14 (Maschke). *Let G be a finite group. Then $\mathbf{F}(G)$ is a semisimple ring if and only if* $\mathrm{char}(\mathbf{F}) = 0$ *or* $\mathrm{char}(\mathbf{F})$ *is relatively prime to the order of G.*

Proof. First, consider the case where $\mathrm{char}(\mathbf{F}) = 0$ or is relatively prime to the order of G. By Theorems 2.2.7 and 2.2.12, it suffices to show that every submodule M_1 of an arbitrary $\mathbf{F}(G)$-module M is complemented.

Let $\iota : M_1 \to M$ be the inclusion. We will construct $\pi : M \to M_1$ with $\pi\iota = 1_{M_1}$. Assuming that, we have a split exact sequence

$$0 \longrightarrow M_1 \overset{\iota}{\longrightarrow} M \longrightarrow M/M_1 \longrightarrow 0,$$

so $M = M_1 \oplus \mathrm{Ker}(\pi)$, and hence M_1 is complemented.

Now, M_1 is a vector subspace of M, so as an $\mathbf{F}$-module M_1 is complemented and there is certainly a linear map $\rho : M \to M_1$ with $\rho\iota = 1_{M_1}$. This is an $\mathbf{F}$-module homomorphism, but there is no reason to expect it to be an $\mathbf{F}(G)$-module homomorphism. We obtain one by averaging it. (Since we are dealing with a single representation here, for simplicity we will write $g(v)$ instead of $\sigma(g)(v)$.)

Let $\pi = \mathrm{Av}(\rho)$. By Lemma 1.13, $\pi \in \mathrm{Hom}_G(M, M_1)$. We have $\rho\iota = 1_{M_1}$; we need $\pi\iota = 1_{M_1}$. Let $v \in M_1$. Then, since M_1 is an $\mathbf{F}(G)$-submodule,

$g(v) \in M_1$ for all $g \in G$, so $\rho(\iota(g(v))) = g(v)$ for all $g \in G$. Then

$$
\begin{aligned}
\pi(\iota(v)) &= \frac{1}{n}\sum_{g \in G} g^{-1}\rho(g(\iota(v))) \\
&= \frac{1}{n}\sum_{g \in G} g^{-1}\rho(\iota(g(v))) \\
&= \frac{1}{n}\sum_{g \in G} g^{-1}(g(v)) \\
&= \frac{1}{n}\sum_{g \in G} v \\
&= \frac{nv}{n} \\
&= v
\end{aligned}
$$

as required.

Now suppose that $\mathrm{char}(\mathbf{F})$ divides the order of G. Recall that we have the augmentation map ε as defined in Example 1.4 (11), giving a short exact sequence of $\mathbf{F}(G)$-modules

$$(1.5) \qquad\qquad 0 \longrightarrow \mathcal{R}_0 \longrightarrow \mathcal{R} \xrightarrow{\;\varepsilon\;} \tau \longrightarrow 0.$$

Suppose that $\mathcal{R}$ were semisimple. Then ε would have a splitting α, $\mathcal{R} \cong \mathcal{R}_0 \oplus \alpha(\tau)$. Since this is a direct sum, $\mathcal{R}_0 \cap \alpha(\tau) = \{0\}$. On the other hand, $\alpha(\tau)$ is a trivial subrepresentation in $\mathcal{R}$, so by the proof of Lemma 1.10,

$$\alpha(\tau) = \left\{ a\sum_{g \in G} g : a \in \mathbf{F}\right\}.$$

However,

$$\varepsilon\left(a\sum_{g \in G} g\right) = a\varepsilon\left(\sum_{g \in G} g\right) = an = 0 \in \mathbf{F},$$

since $\mathrm{char}(\mathbf{F})$ divides n. Thus, $\alpha(\tau) \subseteq \mathcal{R}_0$, contradicting $\mathcal{R}_0 \cap \alpha(\tau) = \{0\}$.
$\square$

Example 1.15. Consider Example 1.4 (10) again, but this time with $\mathbf{F} = \mathbf{F}_p$, the field of p elements. Then $T^p = \begin{bmatrix} 1 & p \\ 0 & 1 \end{bmatrix} = \begin{bmatrix} 1 & 0 \\ 0 & 1 \end{bmatrix}$, so we may regard T as giving an $\mathbf{F}$-representation of $\mathbb{Z}/p\mathbb{Z}$. As in Example 1.12, this is indecomposable but not irreducible, so is not semisimple.

Lemma 1.16. *Let G be a finite group and $\mathbf{F}$ a good field. Let $\sigma : G \to \mathrm{Aut}(M)$ be an $\mathbf{F}$-representation of G of finite degree. Then for each $g \in G$, the linear transformation $\sigma(g) : M \to M$ is diagonalizable.*

Proof. Since $g \in G$, we have that $g^k = 1$ for some k dividing u. Then the minimum polynomial of $\sigma(g)$ divides $X^k - 1$, and hence $X^u - 1$. Since $\mathbf{F}$ is good, the minimum polynomial of $\sigma(g)$ factors into a product of distinct linear factors, so $\sigma(g)$ is diagonalizable. $\qquad\square$

3.2. Representations of Abelian Groups

Before developing the general theory, we will first directly develop the representation theory of finite abelian groups over good fields. We will then see the similarities and differences between the situation for abelian groups and that for groups in general.

Theorem 2.1. *Let G be a finite group and $\mathbf{F}$ a good field for G. Then G is abelian if and only if every irreducible $\mathbf{F}$-representation of G is 1-dimensional.*

Proof. First assume that G is abelian and let M be an $\mathbf{F}$-representation of G. If M is irreducible, then M is isomorphic to a quotient of $\mathcal{R}$, by corollary 2.1.4, so $\deg(M) < \infty$. Now the representation is given by a homomorphism $\sigma : G \to \mathrm{Aut}(M)$, so

$$\sigma(g)\sigma(h) = \sigma(gh) = \sigma(hg) = \sigma(h)\sigma(g) \quad \text{for all } g, h \in G.$$

By Lemma 1.16, each $\sigma(g)$ is diagonalizable, so

$$\mathcal{S} = \{\sigma(g) \mid g \in G\}$$

is a set of mutually commuting diagonalizable transformations.

We now recall a standard, though perhaps not widely appreciated, theorem of linear algebra (see [AW, Theorem 4.3.36] or [HK, Section 6.5]): *Let V be a finite-dimensional vector space over a field $\mathbf{F}$ and let $S = \{s_i : V \to V\}_{i \in I}$ be a set of mutually commuting diagonalizable linear transformations. Then S is simultaneously diagonalizable.* (Outline of proof: Since each s_i is diagonalizable, for each i, V is the direct sum of the eigenspaces of s_i. Furthermore, since the $\{s_i\}$ commute, s_j leaves any eigenspace of s_i invariant for each $i, j \in I$, as if v is an eigenvector of s_i with eigenvalue λ, i.e., $s_i(v) = \lambda v$, then $s_i s_j(v) = s_j s_i(v) = s_j(\lambda v) = \lambda s_j(v)$, so $s_j(v)$ is also an eigenvector of s_i with eigenvalue λ. Now argue by induction on the dimension d of V. If $d = 1$, i.e., V is 1-dimensional, the result is automatic. Assume it is true for all vector spaces of dimension less than d, and let V have dimension d. If every element of S is a scalar multiple of the identity, the result is automatic, so assume not. Then some s_i has more than one eigenvalue, so V has a direct sum decomposition $V = V_1 \oplus \cdots \oplus V_k$ into eigenspaces of s_i, with $k > 1$, and then each summand has dimension less than d, so by induction we are done.)

We apply this theorem here with $S = \mathcal{S}$. If $\mathcal{B} = \{v_1, \ldots, v_k\}$ is a basis of M in which all the elements of $\mathcal{S}$ are diagonal, then

$$M \cong \mathbf{F}v_1 \oplus \mathbf{F}v_2 \oplus \cdots \oplus \mathbf{F}v_k$$

(thus showing directly that M is semisimple). Hence, M is simple if and only if $k = 1$, i.e., if and only if $\deg(M) = 1$.

Conversely, let M be a 1-dimensional representation of G. Then this representation is defined by a homomorphism $\sigma : G \to \operatorname{Aut}(M) \cong \mathbf{F}^*$, and $\mathbf{F}^*$ is, of course, abelian, so that

$$(2.1) \qquad\qquad \sigma(g)\sigma(h) = \sigma(h)\sigma(g) \quad \text{for all } g, h \in G.$$

By assumption, every irreducible representation of G is 1-dimensional, so equation (2.1) is valid for every irreducible representation of G. Since every representation of G is a direct sum of irreducible representations, equation (2.1) is valid for every representation of G. In particular, it is valid for $\mathcal{R}$, the regular representation of G. If σ_0 is the homomorphism defining the regular representation, and we consider $1 \in \mathcal{R}$, then for any $g, h \in G$,

$$(\sigma_0(g)\sigma_0(h))(1) = (\sigma_0(h)\sigma_0(g))(1),$$
$$\sigma_0(g)(h) = \sigma_0(h)(g),$$
$$gh = hg,$$

and G is abelian. $\qquad\qquad\qquad\qquad\qquad\qquad\qquad\qquad\qquad\qquad\qquad\square$

Theorem 2.2. *Let G be a finite group and $\mathbf{F}$ a good field for G. Then G is abelian if and only if G has n distinct irreducible $\mathbf{F}$-representations.*

Proof. Let G be abelian. We shall construct n distinct $\mathbf{F}$-representations of G. We know that we may write G as a direct sum of cyclic groups

$$(2.2) \qquad\qquad G \cong \mathbb{Z}/n_1\mathbb{Z} \oplus \mathbb{Z}/n_2\mathbb{Z} \oplus \cdots \oplus \mathbb{Z}/n_s\mathbb{Z}$$

with $n = n_1 \cdots n_s$. Since $\mathbf{F}$ is good for G, it is also good for each of the cyclic groups $\mathbb{Z}/n_i\mathbb{Z}$, and by Example 1.4 (6), the cyclic group $\mathbb{Z}/n_i\mathbb{Z}$ has the n_i distinct $\mathbf{F}$-representations θ_k for $0 \leq k \leq n_i - 1$; to distinguish these representations for different i, we shall denote them $\theta_k^{n_i}$. Thus, $\theta_k^{n_i} : \mathbb{Z}/n_i\mathbb{Z} \to \mathbf{F}^*$. If $\pi_i : G \to \mathbb{Z}/n_i\mathbb{Z}$ denotes the projection, then

$$(2.3) \qquad\qquad \theta_k^{n_i}\pi_i : G \to \mathbf{F}^* = \operatorname{Aut}(\mathbf{F})$$

defines a 1-dimensional representation (and, hence, an irreducible representation) of G. Thus,

$$\{\theta_{k_1}^{n_1}\pi_1 \otimes \cdots \otimes \theta_{k_s}^{n_s}\pi_s \mid 0 \le k_i \le n_i - 1, \ i = 1, \cdots s\}$$

is a collection of n 1-dimensional, and hence irreducible, $\mathbf{F}$-representations of G. (More simply stated, $\theta_{k_1}^{n_1}\pi_1 \otimes \cdots \otimes \theta_{k_s}^{n_s}\pi_s : G \to \mathbf{F}^* = \mathrm{Aut}(\mathbf{F})$ by $(\theta_{k_1}^{n_1}\pi_1 \otimes \cdots \otimes \theta_{k_s}^{n_s}\pi_s)(g) = \theta_{k_1}^{n_1}\pi_1(g) \cdots \theta_{k_s}^{n_s}\pi_s(g)$ for every $g \in G$.) By Theorem 2.2.14, each $\mathbf{F}$-representation of G appears in $\mathcal{R}$, and $\dim_{\mathbf{F}} \mathcal{R} = n$, so there cannot be any others.

On the other hand, suppose that G is not abelian, and let $\{M_i\}_{i\in I}$ be the set of irreducible representations of G. Since G is not abelian, Theorem 2.1 implies that $\deg(M_i) > 1$ for some i. Then, again by Theorem 2.2.14,

$$|I| = \sum_{i\in I} 1 < \sum_{i\in I} \deg(M_i) \le \deg(\mathbf{F}(G)) = n,$$

so $|I| < n$, as claimed. $\qquad\square$

Corollary 2.3. *Let G be a finite abelian group and $\mathbf{F}$ a good field for G. If $M_1, \ldots, M_n$ denote the distinct irreducible $\mathbf{F}$-representations of G, then*

$$\mathbf{F}(G) = \bigoplus_{i=1}^{n} M_i.$$

(In other words, every irreducible representation of G appears in the regular representation with multiplicity one.)

Proof. If $M = \bigoplus_{i=1}^{n} M_i$, then, by Theorem 2.2.14, M is a subrepresentation of $\mathcal{R}$. But $\deg(M) = n = \deg(\mathcal{R})$, so $M = \mathcal{R}$. $\qquad\square$

Corollary 2.4. *Let G be a finite abelian group and $\mathbf{F}$ a good field for G. If M is an irreducible representation of G, then*

$$\mathrm{End}_G(M) = \mathbf{F}.$$

Proof. Every $\mathbf{F}(G)$-homomorphism is an $\mathbf{F}$-homomorphism and M is 1-dimensional, so $\mathbf{F} \subseteq \mathrm{End}_G(M) \subseteq \mathrm{End}(M) = \mathbf{F}$. $\qquad\square$

Proposition 2.5. *Let G be a group and suppose that $Q = G/[G : G]$ is finite. Let $\mathbf{F}$ be a good field for Q. Then G has $|Q|$ distinct 1-dimensional $\mathbf{F}$-representations.*

Proof. If π is the canonical projection $\pi : G \to Q$, then for any 1-dimensional representation $\sigma : Q \to \mathrm{Aut}(\mathbf{F})$, its pullback $\pi^*(\sigma) = \sigma\pi$ is a 1-dimensional representation of G, and by Theorem 2.2, we obtain $|Q|$ distinct representations in this way.

On the other hand, if $\sigma' : G \to \mathrm{Aut}(\mathbf{F}) = \mathbf{F}^*$ is any 1-dimensional representation of G, then $\sigma'|[G : G]$ is trivial (as $\mathbf{F}^*$ is abelian). Thus, σ' factors as $\sigma\pi$, so it is one of the representations constructed in the last paragraph. $\qquad\square$

3.3. Decomposition of the Regular Representation

Definition 3.1. Let G be a finite group. A field $\mathbf{F}$ is called **excellent for** G or simply **excellent** if it is of characteristic zero or prime to the order of G and is a splitting field for $R = \mathbf{F}(G)$.

The condition on the characteristic of $\mathbf{F}$ ensures that R is semisimple. Recall from Corollary 2.4.12 that if $\mathbf{F}$ is algebraically closed, then $\mathbf{F}$ is certainly a splitting field for R. Thus, in particular, the field $\mathbf{C}$ is excellent for every G. Our objective in this section is to count the number of irreducible representations of a finite group G over an excellent field $\mathbf{F}$, and to determine their multiplicities in the regular representation $\mathbf{F}(G)$. The answers turn out to be both simple and extremely useful.

Lemma 3.2. *Let G be a finite group and $\mathbf{F}$ an arbitrary field. Then* $\dim_{\mathbf{F}} \mathrm{Center}(R) = t$, *the number of conjugacy classes of elements of G.*

Proof. Let $\mathcal{C}$ denote the center of R. Let $\{C_i\}_{i=1}^{t}$ be the conjugacy classes of G, i.e., the sets of mutually conjugate elements of G. Since conjugacy is an equivalence relation, $\{C_i\}_{i=1}^{t}$ is a partition of G. For $1 \leq i \leq t$, let $c_i \in R$ be defined by

$$c_i = \sum_{g \in C_i} g.$$

We claim that

$$\mathcal{C} = \langle c_1, \ldots, c_t \rangle,$$

the $\mathbf{F}$-vector space spanned by these elements. Since $\{c_1, \ldots, c_t\}$ is clearly an $\mathbf{F}$-linearly independent set of elements of R, this implies that

$$\dim_{\mathbf{F}}(\mathcal{C}) = t,$$

as required.

We now prove this claim. For each element g of G, we have the following equality in R:

$$c_i g = \left(\sum_{g_i \in C_i} g_i \right) g = \sum_{g_i \in C_i} g_i g = \sum_{g_i \in C_i} g(g^{-1} g_i g)$$

$$= \sum_{g_i' \in C_i} g g_i' = g \left(\sum_{g_i' \in C_i} g_i' \right) = g c_i,$$

where the fourth equality holds because C_i is a conjugacy class. This immediately implies that $\mathcal{C} \supseteq \langle c_1, \dots, c_t \rangle$.

On the other hand, suppose that we have an element

$$x = \sum_{g \in G} a_g g \in \mathcal{C}.$$

Then for any $g_0 \in G$, we have $g_0 x g_0^{-1} = x g_0 g_0^{-1} = x$. But

$$g_0 x g_0^{-1} = \sum_{g \in G} a_g g_0 g g_0^{-1} = \sum_{g' \in G} a_{g_0^{-1} g' g_0} g',$$

which implies that

$$a_g = a_{g_0^{-1} g g_0} \qquad \text{for all } g, g_0 \in G.$$

That is, any two mutually conjugate elements have the same coefficient. Hence $\mathcal{C} \subseteq \langle c_1, \dots, c_t \rangle$, and we are done. $\qquad\square$

The following theorem, proved by Frobenius in 1896, is fundamental.

Theorem 3.3 (Frobenius). *Let $\mathbf{F}$ be an excellent field for the finite group G.*

(1) *The number of distinct irreducible $\mathbf{F}$-representations of G is equal to the number t of distinct conjugacy classes of elements of G.*

(2) *If $\{M_\alpha\}_{\alpha \in A}$ is a set of representatives of the distinct isomorphism classes of irreducible $\mathbf{F}$-representations of G, then the multiplicity of M_α in the regular representation $\mathcal{R}$ of G is equal to its degree $d_\alpha = \deg(M_\alpha)$, for each $\alpha \in A$.*

(3) $\sum_{\alpha \in A} d_\alpha^2 = n = |G|$.

Proof. Immediate from Lemma 3.2 and Theorem 2.5.1 (which generalized Corollary 2.2.24). $\qquad\square$

This was one of the motivations for our development of semisimple ring theory—to show how the fundamental theorem of group representation theory is a direct consequence of the structure theorems for semisimple rings.

Remark 3.4. Henceforth, we denote the degrees of the irreducible **F**-representations of G, for **F** an excellent field, by $d_1, \dots, d_t$ (instead of $\{d_\alpha\}$) and make the following observations:

(1) $\sum_{i=1}^{t} d_i^2 = n$,

(2) $d_1 = 1$,

(3) $|\{i \mid d_i = 1\}| = $ order of $G/[G:G]$.

Now (1) is just part (3) of Theorem 3.3, (2) follows as G has the representation τ, so some $d_i = 1$, and we number that one d_1, and (3) follows from Proposition 2.5. These provide strong arithmetic conditions which often suffice to determine $\{d_i\}$. We will derive a fourth condition in Theorem 6.3:

(4) d_i divides n for each i.

Remark 3.5. A priori, we should write $d_i(\mathbf{F})$, as the degrees may depend on the characteristic of **F**. However, we will show in Theorem 6.7 that this is not the case—the degrees are independent of the characteristic (and hence the notation d_i is justified).

Remark 3.6. Note that Theorem 3.3 generalizes the results of Section 3.2. For a group G is abelian if and only if every conjugacy class of G consists of a single element, in which case there are $n = |G|$ conjugacy classes. Then G has n distinct irreducible **F**-representations, each of degree 1 and appearing with multiplicity 1 in the regular representation (and $n = \sum_{i=1}^{n} 1^2$).

Warning 3.7. Although the number of conjugacy classes of elements of G is equal to the number of distinct irreducible **F**-representations of G (if **F** is an excellent field for G), there is *no* natural one-to-one correspondence between the set of conjugacy classes and the set of irreducible **F**-representations.

We conclude this section by specializing another two general results.

Theorem 3.8 (Burnside). *Let* **F** *be an excellent field for the finite group* G *and let* $\rho : G \to \mathrm{Aut}(V)$ *be an irreducible* **F***-representation of* G. *Then*

$$\{\rho(g) \mid g \in G\}$$

spans $\mathrm{End}_{\mathbf{F}}(V)$.

Proof. Immediate from Corollary 2.5.4. □

Theorem 3.9 (Frobenius-Schur). *Let* $\mathbf{F}$ *be an excellent field for the finite group* G, *and let* $\{M_\alpha\}_{\alpha \in A}$ *be representatives of the distinct irreducible* $\mathbf{F}$-*representations of* G. *Pick bases for each* M_α *and consider the* n *vectors* $\{f_{ij}^\alpha(g)\}$, $g \in G$, *where* $\alpha \in A$, *and* f_{ij}^α *are the matrix entries of* σ_α *in any basis*, $\sigma_\alpha : \mathbf{F}(G) \to \mathrm{End}(M_\alpha)$. *Then these vectors form a basis for* $\mathbf{F}^n$.

Proof. Immediate from Corollary 2.5.7. $\qquad\square$

As an application of Burnside's theorem (Theorem 3.8), we may obtain further information on representations.

Corollary 3.10. *Let* $\mathbf{F}$ *be an excellent field for the finite group* G *and let* $\rho : G \to \mathrm{Aut}(V)$ *be an irreducible* $\mathbf{F}$-*representation of* G. *Let* $Z(G)$ *be the center of* G, $Z(G) = \{g \in G \mid gx = xg \text{ for all } x \in G\}$. *Then for each* $g \in Z(G)$, $\rho(g)$ *is a homothety (i.e., a scalar multiple of the identity).*

Proof. If $g \in Z(g)$, $\rho(g)$ must be in the center of $\mathrm{End}_{\mathbf{F}}(V)$, which consists of scalar matrices. (If ρ has degree d, then choosing a basis gives an isomorphism of $\mathrm{End}_{\mathbf{F}}(V)$ with $\mathrm{Mat}_d(\mathbf{F})$, and an easy computation shows that a matrix which commutes with every matrix in $\mathrm{Mat}_d(\mathbf{F})$ must be a scalar matrix.) $\qquad\square$

Corollary 3.11. *Let* $Z(G)$ *denote the center of* G. *Then for each* i,

$$d_i^2 \leq [G : Z(G)].$$

Proof. If $\rho : G \to \mathrm{Aut}_{\mathbf{F}}(V)$ is an irreducible representation of degree d with $\mathbf{F}$ an excellent field for G, then $\rho(zg) = \rho(z)\rho(g)$ for each z in $Z(G)$ and each g in G, and $\rho(z)$ is multiplication by a scalar. Hence, if $g_1, \ldots, g_k$ are a set of coset representatives for $Z(G)$ in G, $\rho(g_1), \ldots, \rho(g_k)$ span $\mathrm{End}_{\mathbf{F}}(V)$, of dimension d^2, so $d^2 \leq k = [G : Z(G)]$. $\qquad\square$

3.4. Applications of Frobenius's Theorem

In this section we will find all of the representations of several finite groups G over the field of complex numbers $\mathbf{C}$. We will construct them in various ways. However, in each case Frobenius's theorem will tell us that we have found them all (and will often help us by telling us what to look for).

Example 4.1. Consider the dihedral group

$$G = D_{2m} = \langle x, y \mid x^m = 1, y^2 = 1, xy = yx^{-1} \rangle.$$

For m odd, G has the following conjugacy classes:

$$\{1\},\ \{x,\ x^{m-1}\},\ \{x^2,\ x^{m-2}\},\ \dots,\ \{x^{(m-1)/2},\ x^{(m+1)/2}\},$$
$$\{y,\ xy,\ x^2y,\ \dots,\ x^{m-1}y\}.$$

There are $(m+3)/2$ conjugacy classes, and in Example 1.4 (7) we constructed $(m+3)/2$ distinct irreducible representations over $\mathbf{C}$, so by Theorem 3.3 we have found all of them.

For m even, G has the following conjugacy classes:

$$\{1\},\ \{x,\ x^{m-1}\},\ \{x^2,\ x^{m-2}\},\ \dots,\ \{x^{\frac{m}{2}-1},\ x^{\frac{m}{2}+1}\},\ \{x^{\frac{m}{2}}\},$$
$$\{x^i y \mid i \text{ is even}\},\ \{x^i y \mid i \text{ is odd}\}.$$

There are $\frac{m}{2}+3$ of these, and in Example 1.4 (7) we also constructed $\frac{m}{2}+3$ irreducible representations over $\mathbf{C}$, so again we have found all of them.

Note that G has two (irreducible) 1-dimensional representations for m odd, as then $|G/[G:G]| = 2$, and four (irreducible) 1-dimensional representations for m even, as then $|G/[G:G]| = 4$. We have found them: ψ_+ and ψ_- for m odd; ψ_{++}, ψ_{+-}, ψ_{-+}, and ψ_{--} for m even. This gives an easy proof that the representations φ_k, $k \neq 0, m/2$, are irreducible—one needs only to check that no φ contains a ψ as a subrepresentation, and that is easy.

Note also that $2m = 2(1^2) + \big((m-1)/2\big)(2)^2$ for m odd, and $2m = 4(1)^2 + \big(\frac{m}{2}-1\big)(2)^2$ for m even.

Finally, note that the decomposition in Example 1.8 (4) is as predicted by Theorem 3.3.

Example 4.2. Let us construct all irreducible $\mathbf{C}$-representations of the quaternion group $G = Q_8$. Recall that

$$Q_8 = \{\pm\mathbf{1}, \pm\mathbf{i}, \pm\mathbf{j}, \pm\mathbf{k}\},$$

and it is straightforward to compute that it has conjugacy classes

$$\{\mathbf{1}\},\ \{-\mathbf{1}\},\ \{\pm\mathbf{i}\},\ \{\pm\mathbf{j}\},\ \{\pm\mathbf{k}\}.$$

Thus, we have five irreducible representations whose degrees satisfy $\sum_{i=1}^{5} d_i^2 = 8$, which forces $d_1 = d_2 = d_3 = d_4 = 1$ and $d_5 = 2$. (Actually, it is unnecessary here to find the number of conjugacy classes, for the equation $\sum_{i=1}^{t} d_i^2 = 8$ only has the solutions

$$(2,\ 2),\ (1,\ 1,\ 1,\ 1,\ 1,\ 1,\ 1,\ 1),\ \text{and}\ (1,\ 1,\ 1,\ 1,\ 2).$$

The first of these is impossible since we must have $d_1 = 1$, and the second cannot be correct because G is nonabelian, so the third must apply.)

Note that $C = \{\pm 1\}$ is the center of G, and hence a normal subgroup of G, and so we have an exact sequence

$$1 \longrightarrow C \longrightarrow G \xrightarrow{\pi} V \longrightarrow 1$$

where $V \cong \mathbb{Z}/2\mathbb{Z} \oplus \mathbb{Z}/2\mathbb{Z}$. Since V is abelian of order 4, it has four 1-dimensional representations, and their pullbacks give four 1-dimensional (and hence certainly irreducible) representations of G.

To be precise, let

$$V = \langle 1, I, J, K \mid I^2 = J^2 = K^2 = 1,\ IJ = K \rangle.$$

Then $\pi : G \to V$ is defined by $\pi(\pm \mathbf{1}) = 1$, $\pi(\pm \mathbf{i}) = I$, $\pi(\pm \mathbf{j}) = J$, and $\pi(\pm \mathbf{k}) = K$. The representations of V are the trivial representation $\sigma_0 = \tau$ (and $\pi^*(\tau) = \tau$) and the representations σ_i for $i = 1, 2, 3$ given by

$$
\begin{array}{llll}
\sigma_1 : & \sigma_1(I) = 1, & \sigma_1(J) = -1, & \sigma_1(K) = -1, \\
\sigma_2 : & \sigma_2(I) = -1, & \sigma_2(J) = 1, & \sigma_2(K) = -1, \\
\sigma_3 : & \sigma_3(I) = -1, & \sigma_3(J) = -1, & \sigma_3(K) = 1.
\end{array}
$$

We also need to find a 2-dimensional representation ρ of Q_8. Here it is:

$$\rho(\pm \mathbf{1}) = \begin{bmatrix} \pm 1 & 0 \\ 0 & \pm 1 \end{bmatrix},$$

$$\rho(\pm \mathbf{i}) = \begin{bmatrix} \pm i & 0 \\ 0 & \mp i \end{bmatrix},$$

$$\rho(\pm \mathbf{j}) = \begin{bmatrix} 0 & \pm i \\ \pm i & 0 \end{bmatrix},$$

$$\rho(\pm \mathbf{k}) = \begin{bmatrix} 0 & \mp 1 \\ \pm 1 & 0 \end{bmatrix}.$$

Note that in these matrices, i is the complex number i. We must check that ρ is irreducible. This can be done directly, but it is easier to make the following observation. If ρ were not irreducible it would have to be isomorphic to $\pi^*(\sigma_i) \oplus \pi^*(\sigma_j)$ for some $i,\ j \in \{0, 1, 2, 3\}$. However, it cannot be, for $\rho(-1)$ is nontrivial, but $(\pi^*(\sigma_i) \oplus \pi^*(\sigma_j))(-1)$ is trivial for any choice of i and j.

Example 4.3. Let us construct all irreducible **C**-representations of the alternating group A_4 of order 12. Recall that A_4 is the subgroup of the symmetric group S_4 consisting of the even permutations, and it is a semi-direct product

$$1 \longrightarrow V \longrightarrow A_4 \xrightarrow{\pi} S \longrightarrow 1,$$

where

$$V \cong \mathbb{Z}/2\mathbb{Z} \oplus \mathbb{Z}/2\mathbb{Z}$$
$$= \{1,\ (1\ 2)(3\ 4),\ (1\ 3)(2\ 4),\ (1\ 4)(2\ 3)\}$$
$$= \{1,\ I,\ J,\ K\}$$

and
$$S \cong \mathbb{Z}/3\mathbb{Z} = \{1,\ (1\ 2\ 3),\ (1\ 3\ 2)\} = \{1,\ T,\ T^2\}.$$

We compute that A_4 has four conjugacy classes

$$\{1\},\ \{I,\ J,\ K\},\ \{T,\ TI,\ TJ,\ TK\},\ \{T^2,\ T^2I,\ T^2J,\ T^2K\},$$

so we expect to find four irreducible representations whose degrees satisfy $\sum_{i=1}^{4} d_i^2 = 12$, giving $d_1 = d_2 = d_3 = 1$ and $d_4 = 3$. (Alternatively, we find that V is the commutator subgroup of G, and so we have exactly three 1-dimensional representations of G, by Proposition 2.5. Then the equation $\sum_{i=1}^{t} d_i^2 = 12$ and $d_1 = d_2 = d_3 = 1$ with $d_i > 1$ for $i > 3$ forces $t = 4$ and $d_4 = 3$.)

The three 1-dimensional representations of G are $\pi^*(\theta_i)$ for $0 \le i \le 2$, where θ_i are the representations of the cyclic group S constructed in Example 1.4 (6).

Now we need to find a 3-dimensional representation. Let

$$M = \mathbf{C}^4 = \{[z_1,\ z_2,\ z_3,\ z_4] \mid z_i \in \mathbf{C},\ 1 \le i \le 4\}.$$

(Recall our convention that $[*, \ldots , *]$ denotes a column vector.) Then S_4, and hence, A_4, acts on $\mathbf{C}^4$ by permuting the coordinates, i.e.,

$$g([z_1, z_2, z_3, z_4]) = [z_{g(1)}, z_{g(2)}, z_{g(3)}, z_{g(4)}] \quad \text{for } g \in S_4.$$

Consider
$$M_0 = \{[z_1, z_2, z_3, z_4] \mid z_1 + z_2 + z_3 + z_4 = 0\}.$$

This subspace of M is invariant under S_4, so it gives a 3-dimensional representation α of S_4, and we consider its restriction to A_4, which we still denote by α. We claim that this representation is irreducible, and the argument is the same as the final observation in Example 4.2: The subgroup V acts trivially in each of the representations $\pi^*(\theta_i)$ for $i = 0, 1, 2$, but nontrivially in the representation α.

Example 4.4. Let us construct all irreducible $\mathbf{C}$-representations of the symmetric group S_4. We have a semidirect product

$$1 \longrightarrow V \longrightarrow S_4 \overset{\pi}{\longrightarrow} W \longrightarrow 1,$$

where V is the same as in Example 4.3 and

$$W = \langle T, U : T^3 = U^2 = 1, TU = UT^2 \rangle \cong D_6 \cong S_3,$$

where $T = (1\,2\,3)$ as before and $U = (1\,2)$. Now (see [AW, Corollary 1.5.10]), S_4 has five conjugacy classes, so we look for a solution of $\sum_{i=1}^{5} d_i^2 = 24 = |S_4|$. This has the unique solution

$$d_1 = d_2 = 1, \; d_3 = 2, \; d_4 = d_5 = 3.$$

As before, we obtain an irreducible representation of S_4 from every irreducible representation of W, and by Example 1.4 (7), W has the irreducible representations $\psi_+ = \tau$, ψ_-, and φ_1, so we have irreducible representations $\pi^*(\tau) = \tau$, $\pi^*(\psi_+)$, and $\pi^*(\varphi_1)$ of degrees 1, 1, and 2, respectively.

We need to find two 3-dimensional irreducible representations. For the first we simply take the representation α constructed in Example 4.3. Since the restriction of α to A_4 is irreducible, α itself is certainly irreducible.

For the second, we take $\alpha' = \pi^*(\psi_-) \otimes \alpha$. This is also a 3-dimensional representation of S_4 (which we may now view as acting on M_0 by

$$\alpha'(g)(v) = \psi_-\pi(g)\alpha(v) \quad \text{for } v \in M_0$$

since $\psi_-\pi(g) = \pm 1$). Since α' restricted to A_4 agrees with α restricted to A_4 and the latter is irreducible, so is the former. To complete our construction, we need only show that α and α' are inequivalent. We see this as $\alpha(U)$ has characteristic polynomial $(X - 1)^2(X + 1)$, while $\alpha'(U)$ has characteristic polynomial $(X - 1)(X + 1)^2$.

(The reader may wonder about the representation $\pi^*(\psi_-) \otimes \pi^*(\varphi_1)$, but this is isomorphic to $\pi^*(\varphi_1)$, so it gives nothing new.)

Remark 4.5. Observe that in all of these examples, all of the representations were defined over a good field.

We conclude this section with a brief look at two examples which illustrate the situation when $\mathbf{F}$ is not algebraically closed.

Example 4.6. We consider the representations of $\mathbb{Z}/p\mathbb{Z}$ over $\mathbf{Q}$. We have already seen in Example 1.7 that $\mathbb{Z}/p\mathbb{Z}$ has two irreducible representations over $\mathbf{Q}$, the representations τ of degree 1 and $\mathcal{R}_0$ of degree $p - 1$, and that $\mathcal{R} \cong \tau \oplus \mathcal{R}_0$. Since R is commutative, $\text{Center}(R) = R$ has dimension p. Now $\text{End}_R(\tau) \cong \mathbf{Q}$ and $\text{End}_R(\mathcal{R}_0) \cong \mathbf{Q}(\sqrt[p]{1})$, an extension of $\mathbf{Q}$ of degree $p - 1$. Looking at Corollary 2.2.24, we verify the equalities:

(1) $p = 1 + (p - 1)$,

(2) $1 = 1 \cdot 1$ and $p - 1 = 1 \cdot (p - 1)$,

(3) $p = 1^2 \cdot 1 + 1^2 \cdot (p - 1)$.

Example 4.7. We consider the representations of Q_8 over $\mathbf{R}$. Then Q_8 has five irreducible representations over $\mathbf{R}$. The four 1-dimensional $\mathbf{C}$-representations $\sigma_1, \ldots, \sigma_4$ of Q_8 are each defined over $\mathbf{R}$, and for each such representation σ, $\mathrm{End}_R(\sigma) \cong \mathbf{R}$. There is a fifth irreducible dimensional representation σ_5 of Q_8 over $\mathbf{R}$, of dimension 4 (and $\mathbf{C} \otimes_{\mathbf{R}} \sigma_5$ is equivalent to 2ρ, where ρ is the irreducible 2-dimensional representation of Q_8 over $\mathbf{C}$). Then $\mathrm{End}_{\mathbf{R}}(\sigma_5) \cong H$, the quaternions, and $\mathrm{Center}(H) = \mathbf{R}$. Of course, $\mathrm{Center}(R)$ has dimension 5 as Q_8 has five conjugacy classes. Looking at Corollary 2.2.24, we verify the equalities:

(1) $5 = 1 + 1 + 1 + 1 + 1$,

(2) $1 = 1 \cdot 1$ (4 times) and $4 = 1 \cdot 4$,

(3) $8 = 1^2 \cdot 1 + 1^2 \cdot 1 + 1^2 \cdot 1 + 1^2 \cdot 1 + 1^2 \cdot 4$.

3.5. Characters

In this section we develop the theory of characters. In practice, characters are a tool whose usefulness, especially in characteristic zero, can hardly be overemphasized. We will begin without restricting the characteristic. *However, we will assume that all representations are finite dimensional.*

Definition 5.1. Let $\sigma : G \to \mathrm{Aut}(M)$ be an $\mathbf{F}$-representation of G of finite degree, and let $\mathcal{B}$ be a basis of M. The **character** of the representation σ is the function $\chi_\sigma : G \to \mathbf{F}$ defined by

$$(5.1) \qquad \chi_\sigma(g) = \mathrm{Tr}([\sigma(g)]_{\mathcal{B}}).$$

Recall that Tr denotes the trace of a matrix or of a linear transformation, and that this is independent of the choice of the basis $\mathcal{B}$. By the same logic, it is the case that if two representations are equivalent, then their characters are equal. It is one of the great uses of characters that, under the proper circumstances, the converse of this is true as well.

Example 5.2.

(1) If $\sigma = d\tau$, then $\chi_\sigma(g) = d$ for every $g \in G$.

(2) If σ is any representation of degree d, then $\chi_\sigma(1) = d$.

(3) If σ is the regular representation of G, then $\chi_\sigma(1) = n = |G|$ and $\chi_\sigma(g) = 0$ for all $g \neq 1$. (To see this, consider $[\sigma(g)]$ in the basis $\{g \mid g \in G\}$ of $\mathbf{F}(G)$.)

Lemma 5.3. *If g_1 and g_2 are conjugate elements of G, then for any representation σ,*

$$(5.2) \qquad\qquad \chi_\sigma(g_1) = \chi_\sigma(g_2).$$

Proof. If $g_2 = g g_1 g^{-1}$, then

$$
\begin{aligned}
\chi_\sigma(g_2) = \chi_\sigma(g g_1 g^{-1}) &= \mathrm{Tr}\left([\sigma(g g_1 g^{-1})]_\mathcal{B}\right) \\
&= \mathrm{Tr}\left([\sigma(g)]_\mathcal{B}[\sigma(g_1)]_\mathcal{B}[\sigma(g)]_\mathcal{B}^{-1}\right) = \mathrm{Tr}\left([\sigma(g_1)]_\mathcal{B}\right) \\
&= \chi_\sigma(g_1).
\end{aligned}
$$

$\square$

Proposition 5.4. *Let σ_1 and σ_2 be representations of G. Then*

(1) $\chi_{\sigma_1 \oplus \sigma_2} = \chi_{\sigma_1} + \chi_{\sigma_2}$, *and*

(2) $\chi_{\sigma_1 \otimes \sigma_2} = \chi_{\sigma_1} \chi_{\sigma_2}$.

Proof. (1) is obvious, and (2) follows immediately from the matrix representation of the tensor product (see [AW, Lemma 4.1.20 and Proposition 7.2.35]). $\square$

Proposition 5.5.

(1) *Let $\mathrm{char}(\mathbf{F}) = p \neq 0$, and let σ be an $\mathbf{F}$-representation of G. Then for every $g \in G$, $\chi_\sigma(g)$ is an element of the field $\mathbf{F}_p(\sqrt[u]{1})$.*

(2) *Let $\mathrm{char}(\mathbf{F}) = 0$, and let σ be an $\mathbf{F}$-representation of G. Then for any $g \in G$, $\chi_\sigma(g)$ is an algebraic integer in the field $\mathbf{Q}(\sqrt[u]{1})$.*

Proof. In either case, $\sigma(g)$ satisfies the polynomial $X^u - 1$, so over the given field $\sigma(g)$ has a Jordan form in which each diagonal entry is a u-th root of 1. Thus $\chi_\sigma(g)$ is a sum of u-th roots of 1 in that field.

(Recall that an algebraic number is an algebraic integer if it is a root of a monic polynomial with (ordinary) integer coefficients. Since $X^u - 1$ is such a polynomial, every u-th root of 1 is an algebraic integer. Furthermore, the algebraic integers form a ring, so every sum of u-th roots of 1 is an algebraic integer. See [Sa, Section II.8].) $\square$

Our next goal is to derive the basic orthogonality results for characters. Along the way, we will derive a bit more: orthogonality for matrix coefficients.

Proposition 5.6. *Let $\mathbf{F}$ be an excellent field for the group G, let V_i be irreducible representations of G, given by $\sigma_i : G \to \mathrm{Aut}(V_i)$, and let $\mathcal{B}_i$ be a basis of V_i, for $i = 1, 2$.*

For $g \in G$, let

$$P(g) = [p_{i_1 j_1}(g)] = [\sigma_1(g)]_{\mathcal{B}_1}$$

and

$$Q(g) = [q_{i_2 j_2}(g)] = [\sigma_2(g)]_{\mathcal{B}_2}.$$

(1) *Suppose that V_1 and V_2 are distinct. Then for any i_1, j_1, i_2, j_2,*

$$(5.3) \qquad \frac{1}{n} \sum_{g \in G} p_{i_1 j_1}(g) q_{i_2 j_2}(g^{-1}) = 0.$$

(2) *Suppose that $V_1 = V_2$ (so that $\sigma_1 = \sigma_2$) and $\mathcal{B}_1 = \mathcal{B}_2$. Let $d = \deg(V_1)$. Then*

$$(5.4) \qquad \frac{1}{n} \sum_{g \in G} p_{i_1 j_1}(g) q_{i_2 j_2}(g^{-1}) = \begin{cases} 1/d & \text{if } i_1 = j_2 \text{ and } j_1 = i_2, \\ 0 & \text{otherwise.} \end{cases}$$

(Note that in this case $p_{ij}(g) = q_{ij}(g)$, of course.)

Proof. Let β_i be the projection of V_1 onto its i-th summand $\mathbf{F}$ (as determined by the basis $\mathcal{B}_1$) and let α_j be the inclusion of $\mathbf{F}$ onto the j-th summand of V_2 (as determined by the basis $\mathcal{B}_2$). Then $f = \alpha_j \beta_i \in \mathrm{Hom}(V_1, V_2)$. Note that $[\alpha_j \beta_i]_{\mathcal{B}_2}^{\mathcal{B}_1} = E_{ji}$, where E_{ji} is the matrix with 1 in the ji-th position and 0 elsewhere. Let us compute $\mathrm{Av}(f)$. By definition

$$\mathrm{Av}(f) = \frac{1}{n} \sum_{g \in G} \sigma_2(g^{-1})(\alpha_j \beta_i)\sigma_1(g).$$

Direct matrix calculation shows

$$(5.5) \quad [\sigma_2(g^{-1})(\alpha_j \beta_i)\sigma_1(g)]_{\mathcal{B}_2}^{\mathcal{B}_1} = \begin{bmatrix} p_{1j}(g)q_{i1}(g^{-1}) & p_{1j}(g)q_{i2}(g^{-1}) & \cdots \\ p_{2j}(g)q_{i1}(g^{-1}) & p_{2j}(g)q_{i2}(g^{-1}) & \cdots \\ \vdots & \vdots & \ddots \end{bmatrix},$$

so the sums in question are just the entries of $[\mathrm{Av}(f)]_{\mathcal{B}_2}^{\mathcal{B}_1}$ (as we vary i, j and the entry of the matrix).

Consider case (1). Here $\mathrm{Av}(f) \in \mathrm{Hom}_G(V_1, V_2) = \{0\}$ by Schur's lemma, so f is the zero map and all matrix entries are 0, as claimed.

Now for case (2). Here $\mathrm{Av}(f) \in \mathrm{Hom}_G(V,\, V) = \mathbf{F}$, by Schur's lemma, with every element a homothety, represented by a scalar matrix. Thus all the off-diagonal entries of $[\mathrm{Av}(f)]_{\mathcal{B}_1}$ are 0, showing that the sum in equation (5.4) is zero if $i_1 \neq j_2$. Since $\sigma_1 = \sigma_2$, we may rewrite the sum (replacing g by g^{-1}) as

$$\frac{1}{n} \sum_{g \in G} p_{i_2 j_2}(g) q_{i_1 j_1}(g^{-1}),$$

showing that it is 0 if $j_1 \neq i_2$.

Consider the remaining situation, where $i_1 = j_2 = i$ and $j_1 = i_2 = j$. As $\mathrm{Av}(f)$ is a homothety, all of the diagonal entries of its matrix are equal, so we obtain a common value, say x, for

$$\frac{1}{n} \sum_{g \in G} p_{ij}(g) q_{ji}(g^{-1})$$

for any i, j (by varying the choice of f and the diagonal element in question).

For example, letting $f = \alpha_1 \beta_1$ and looking at the first two diagonal entries of $[\mathrm{Av}(f)]_{\mathcal{B}_1}$, we see that

$$\frac{1}{n} \sum_{g \in G} p_{11}(g) q_{11}(g^{-1}) = \frac{1}{n} \sum_{g \in G} p_{21}(g) q_{12}(g^{-1}).$$

Letting $f = \alpha_2 \beta_2$ and looking at the first two diagonal entries of $[\mathrm{Av}(f)]_{\mathcal{B}_1}$, we see as well that

$$\frac{1}{n} \sum_{g \in G} p_{12}(g) q_{21}(g^{-1}) = \frac{1}{n} \sum_{g \in G} p_{22}(g) q_{22}(g^{-1}).$$

But also, since $p_{ij} = q_{ij}$, and as g runs over G so does g^{-1},

$$\frac{1}{n} \sum_{g \in G} p_{12}(g) q_{21}(g^{-1}) = \frac{1}{n} \sum_{g \in G} p_{21}(g) q_{12}(g^{-1}),$$

so all four of these sums are equal.

Now consider

$$f_0 = \alpha_1 \beta_1 + \alpha_2 \beta_2 + \cdots + \alpha_d \beta_d.$$

Since there are d summands, we see that the diagonal entries of $[\mathrm{Av}(f_0)]_{\mathcal{B}_1}$ are all equal to dx. But f_0 is the identity! Hence, $\mathrm{Av}(f_0) = f_0$ and $[\mathrm{Av}(f_0)]_{\mathcal{B}_1}$ is the identity matrix, with its diagonal entries equal to 1, so $dx = 1$ and $x = 1/d$, as claimed. $\qquad\square$

Corollary 5.7. *Let* **F** *be an excellent field for G with* $\mathrm{char}(\mathbf{F}) = p \neq 0$, *and let V be an irreducible* **F**-*representation of G. Then* $\deg(V)$ *is not divisible by p.*

Proof. If $d = \deg(V)$, then the above proof shows that $dx = 1 \in \mathbf{F}$, so that $d \neq 0 \in \mathbf{F}$. $\qquad\square$

Corollary 5.8 (Orthogonality of characters). *Let* **F** *be an excellent field for G, and let V_1 and V_2 be irreducible representations of G defined by* $\sigma_i : G \to \mathrm{Aut}(V_i)$ *for $i = 1,\ 2$. Then*

$$\frac{1}{n}\sum_{g \in G} \chi_{\sigma_1}(g)\chi_{\sigma_2}(g^{-1}) = \begin{cases} 0 & \text{if } V_1 \text{ and } V_2 \text{ are distinct,} \\ 1 & \text{if } V_1 \text{ and } V_2 \text{ are isomorphic.} \end{cases}$$

Proof. If V_i has degree d_i, this sum is equal to

$$\frac{1}{n}\sum_{g \in G}\left(\sum_{i=1}^{d_1}\sum_{j=1}^{d_2} p_{ii}(g)q_{jj}(g^{-1})\right) = \sum_{i=1}^{d_1}\sum_{j=1}^{d_2}\left(\frac{1}{n}\sum_{g \in G} p_{ii}(g)q_{jj}(g^{-1})\right),$$

which is 0 if V_1 and V_2 are distinct. If V_1 and V_2 are isomorphic of degree d, then, since isomorphic representations have the same character, we may assume that $V_1 = V_2$. The terms with $i \neq j$ are all zero, so the sum is

$$\sum_{j=1}^{d} p_{jj}(g)q_{jj}(g^{-1}) = d(1/d) = 1.$$

$$\square$$

Proposition 5.6 and Corollary 5.8 have the following generalization.

Proposition 5.9. *Given the same hypothesis and notation as in Proposition 5.6, let h be a fixed element of G.*

(1) *Suppose that V_1 and V_2 are distinct. Then for any i_1, j_1, i_2, and j_2, we have*

$$\frac{1}{n}\sum_{g \in G} p_{i_1 j_1}(hg)q_{i_2 j_2}(g^{-1}) = 0.$$

(2) *Suppose that $V_1 = V_2$. Then if $i_1 \neq j_2$,*

$$\frac{1}{n}\sum_{g \in G} p_{i_1 j_1}(hg)q_{i_2 j_2}(g^{-1}) = 0.$$

(3) *For any two representations V_1 and V_2,*

$$\frac{1}{n}\sum_{g\in G}\chi_{\sigma_1}(hg)\chi_{\sigma_2}(g^{-1}) = \begin{cases} 0 & \text{if } V_1 \text{ and } V_2 \text{ are distinct,} \\ (1/d)\chi_{\sigma_1}(h) & \text{if } V_1 = V_2 \text{ is of degree d.} \end{cases}$$

Proof. The proof follows that of Proposition 5.6, with $f = \alpha_j\beta_i\sigma_1(h)$. Then the matrix corresponding to the one in equation (5.5) is

(5.6)
$$[\sigma_2(g^{-1})(\alpha_j\beta_i)\sigma_1(hg)]_{\mathcal{B}_2}^{\mathcal{B}_1} = \begin{bmatrix} p_{1j}(hg)q_{i1}(g^{-1}) & p_{1j}(hg)q_{i2}(g^{-1}) & \cdots \\ p_{2j}(hg)q_{i1}(g^{-1}) & p_{2j}(hg)q_{i2}(g^{-1}) & \cdots \\ \vdots & \vdots & \ddots \end{bmatrix}.$$

First suppose that V_1 and V_2 are distinct. Then, again, $\mathrm{Av}(f) = 0$, so all the matrix entries are zero, yielding (1), and then the first assertion of (3) follows as in the proof of Corollary 5.8.

Now suppose that $V_1 = V_2$. Then, again, $\mathrm{Av}(f)$ is a homothety, so the off-diagonal matrix entries are zero, yielding (2), while all the diagonal entries are equal. Set $i = j$ and call the common value in this case x_j, i.e.,

$$x_j = \frac{1}{n}\sum_{g\in G}p_{ij}(hg)q_{ji}(g^{-1}),$$

this sum being independent of i. As in the proof of Corollary 5.8, the sum we are interested in is

$$\sum_{i=1}^{d}\sum_{j=1}^{d}\left(\frac{1}{n}\sum_{g\in G}p_{ii}(hg)q_{jj}(g^{-1})\right) - \sum_{j=1}^{d}\left(\frac{1}{n}\sum_{g\in G}p_{jj}(hg)q_{jj}(g^{-1})\right)$$
$$= x_1 + \cdots + x_d$$

where the first equality follows from part (2).

Now consider
$$f_0 = (\alpha_1\beta_1 + \cdots + \alpha_d\beta_d)\sigma_1(h).$$

Then $\mathrm{Av}(f_0)$ is a homothety, and all of its diagonal entries are equal to $x_1 + \cdots + x_d$, so
$$\mathrm{Tr}(\mathrm{Av}(f_0)) = d(x_1 + \cdots + x_d).$$

But $f_0 = \sigma_1(h)$, so
$$\mathrm{Av}(f_0) = \frac{1}{n}\sum_{g\in G}\sigma_1(g^{-1}hg).$$

Then

$$d(x_1 + \cdots + x_d) = \mathrm{Tr}(\mathrm{Av}(f_0))$$
$$= \frac{1}{n} \sum_{g \in G} \mathrm{Tr}(\sigma_1(g^{-1}hg))$$
$$= \frac{1}{n} \sum_{g \in G} \mathrm{Tr}(\sigma_1(h))$$
$$= \mathrm{Tr}(\sigma_1(h))$$
$$= \chi_{\sigma_1}(h),$$

yielding the desired equality. $\qquad\square$

Warning 5.10.

(1) We should caution the reader that there are in fact three alternatives in Proposition 5.6: that V_1 and V_2 are distinct, that they are equal, or that they are isomorphic but unequal. In this last case the sum in equation (5.4) and the corresponding sum in the statement of Proposition 5.9 may vary. Note that we legitimately reduced the third case to the second in the proof of Corollary 5.8; the point there is that while the individual matrix entries of isomorphic representations will differ, their traces will be the same.

(2) We should caution the reader that the sum in Proposition 5.9 (2) with $i_1 = j_2$ but $j_1 \neq i_2$ may well be nonzero, and that the quantities $x_1, \ldots, x_d$ in the proof of that proposition may well be unequal.

Given a finite group G and an excellent field $\mathbf{F}$, let G have conjugacy classes $C_1, \ldots, C_t$ (in some order) and irreducible representations $\sigma_1, \ldots, \sigma_t$ (in some order).

Definition 5.11. The **character table of** G is the matrix $A \in \mathrm{Mat}_t(\mathbf{F})$ defined by $A = [a_{ij}] = [\chi_i(c_j)]$, where $\chi_i = \chi_{\sigma_i}$ and $c_j \in C_j$.

Let $|C_i|$ be the number of elements of C_i. Then Corollary 5.8 gives a sort of orthogonality relation on the rows of A; namely,

$$(5.7) \qquad \frac{1}{n} \sum_{g \in G} \chi_i(g)\chi_j(g^{-1}) = \frac{1}{n} \sum_{k=1}^{t} |C_k|\chi_i(c_k)\chi_j(c_k)^{-1} = \delta_{ij}.$$

(Recall that $\delta_{ij} = 1$ if $i = j$ and 0 if not.)

From this, we can immediately write down $B = A^{-1}$.

Lemma 5.12. $B = [b_{ij}] = \left[\dfrac{|C_i|}{n} \chi_j(c_i^{-1}) \right].$

Proof. The equation $AB = I$ is the equation $\sum_{k=1}^{t} a_{ik}b_{kj} = \delta_{ij}$, which is immediate from equation (5.7). $\qquad\square$

Now let us interpret B.

Definition 5.13. Let $f : G \to \mathbf{F}$ be a function with the property that $f(g_1) = f(g_2)$ whenever g_1 and $g_2 \in G$ are conjugate. Then f is called a **class function** on G.

Clearly, $f : G \to \mathbf{F}$ is a class function if and only if it is constant on each conjugacy class C_i $(1 \le i \le t)$. Thus the space of class functions is clearly an $\mathbf{F}$-vector space of dimension t with basis $\mathcal{B} = \{f_1, \ldots, f_t\}$, where

$$f_i(g) = \begin{cases} 1 & \text{if } g \in C_i, \\ 0 & \text{otherwise.} \end{cases}$$

Lemma 5.14. *Let $\mathbf{F}$ be an excellent field for G. Then*

$$\mathcal{A} = \{\chi_1, \ldots, \chi_t\}$$

is a basis for the space of class functions on G.

Proof. By Lemma 5.3, the characters are class functions. There are t of them, so to show that they are a basis it suffices to show that they are linearly independent. Suppose

$$a_1\chi_1 + a_2\chi_2 + \cdots + a_t\chi_t = 0 \quad \text{with } a_i \in \mathbf{F},$$

where, as in Definition 5.11, we have written χ_i for χ_{σ_i}. Note that χ_i^*, defined by $\chi_i^*(g) = \chi_i(g^{-1})$, is also a class function. Then for each i

$$(a_1\chi_1 + a_2\chi_2 + \cdots + a_t\chi_t)\chi_i^* = 0,$$

and averaging over the group we see that

$$\frac{1}{n}\sum_{g \in G} a_1\chi_1(g)\chi_i^*(g) + a_2\chi_2(g)\chi_i^*(g) + \cdots + a_t\chi_t(g)\chi_i^*(g) = 0.$$

But, by the orthogonality relations (Corollary 5.8), this sum is just a_i/n, so that $a_i/n = 0$ and $a_i = 0$ for each i, as required. $\qquad\square$

Remark 5.15. If χ is the character of the representation V defined by $\sigma : G \to \mathrm{Aut}(V)$, then $\chi^*(g) = \chi(g^{-1})$ is indeed the character of a representation. With the given action of G, V is a *left* $\mathbf{F}(G)$-module. The action of G given by $g \mapsto \sigma(g^{-1})$ then gives V the structure of a *right* $\mathbf{F}(G)$-module, and then $V^* = \mathrm{Hom}_G(V, \mathbf{F}(G))$ is a *left* $\mathbf{F}(G)$-module with character χ^*. (In terms of matrices, if $\mathcal{B}$ is a basis for V, $\mathcal{B}^*$ is the dual basis of V^*, and χ^* is defined by $\sigma^* : G \to \mathrm{Aut}(V^*)$, then

$$[\sigma^*(g)]_{\mathcal{B}^*} = [\sigma(g^{-1})]_{\mathcal{B}}^t.)$$

Proposition 5.16. *Let* **F** *be an excellent field for* G. *Then* g_1 *and* $g_2 \in$ G *are conjugate if and only if* $\chi_\sigma(g_1) = \chi_\sigma(g_2)$ *for every irreducible* **F***-representation* σ *of* G.

Proof. The only-if part is trivial. As for the if part, suppose that $\chi_\sigma(g_1) = \chi_\sigma(g_2)$ for every irreducible **F**-representation σ. Then, by Lemma 5.14, $f(g_1) = f(g_2)$ for every class function f on G, so in particular $f_i(g_1) = f_i(g_2)$ for each i, where f_i is the function defined following Definition 5.13, and so g_1 and g_2 are conjugate. $\square$

Proposition 5.17. *Let* A *and* B *be the above matrices, and let* χ_i *and* f_i *be defined as above. Then for every* $1 \leq i \leq t$,

(1) $\chi_i = \sum_{j=1}^{t} a_{ij} f_j$, *and*

(2) $f_i = \sum_{j=1}^{t} b_{ij} \chi_j$.

Proof. (1) is easy. We need only show that both sides agree on c_k for $1 \leq k \leq t$. Since $f_j(c_k) = \delta_{jk}$,

$$\sum_{j=1}^{t} a_{ij} f_j(c_k) = a_{ik} = \chi_i(c_k),$$

as claimed. Now (1) says that the change of basis matrix $P_{\mathcal{B}}^{\mathcal{A}}$ from the basis $\mathcal{A}$ to the basis $\mathcal{B}$ is just A. (See [AW, section 4.3] or [HK, section 2.4]). Then

$$P_{\mathcal{A}}^{\mathcal{B}} = \left(P_{\mathcal{B}}^{\mathcal{A}}\right)^{-1} = A^{-1} = B,$$

giving (2). $\square$

As a corollary, we may derive another set of orthogonality relations.

Corollary 5.18. *Let* g_1, $g_2 \in G$. *Then*

$$\sum_{j=1}^{t} \chi_j(g_1)\chi_j(g_2^{-1}) = \begin{cases} 0 & \text{if } g_1 \text{ and } g_2 \text{ are not conjugate,} \\ \dfrac{n}{|C_i|} & \text{if } g_1, g_2 \in C_i. \end{cases}$$

Proof. We may assume that $g_1 = c_i$ and $g_2 = c_k$ for some i, k. Then by the definition of f_k and Proposition 5.16 (2),

$$\delta_{ki} = f_k(c_i) = \left(\sum_{j=1}^{t} b_{kj}\chi_j\right)(c_i)$$

$$= \sum_{j=1}^{t} \left(\frac{|C_k|}{n}\chi_j(c_k^{-1})\right)\chi_j(c_i)$$

$$= \frac{|C_k|}{n}\sum_{j=1}^{t} \chi_j(c_i)\chi_j(c_k^{-1}),$$

yielding the corollary. $\square$

We now define an important quantity.

Definition 5.19. Let G and $\mathbf{F}$ be arbitrary and let V and W be two $\mathbf{F}$-representations of G. The **intertwining number of V and W** is

$$i(V,\ W)\ =\ \langle V,\ W\rangle\ =\ \dim_{\mathbf{F}} \operatorname{Hom}_G(V,\ W).$$

Lemma 5.20. *The intertwining number has the following properties:*

(1) *If $V \cong \bigoplus V_i$ and $W \cong \bigoplus W_j$ with both sums finite, where $\{V_i\}$ and $\{W_j\}$ are $\mathbf{F}$-representations of G, then $i(V,W) = \sum_{i,j} i(V_i, W_j)$.*

(2) *If $\mathbf{E}$ is a field containing $\mathbf{F}$, $V' = \mathbf{E} \otimes_{\mathbf{F}} V$, and $W' = \mathbf{E} \otimes_{\mathbf{F}} W$, then*

$$i(V', W') = i(V, W).$$

Proof. (1) is clear and (2) is immediate from Corollary 2.4.3. $\qquad\square$

In the following lemma, the last equality in (5) is the definition of the intertwining number $\langle \chi_V,\ \chi_W\rangle$ of the characters χ_V and χ_W.

Lemma 5.21. *Let G be finite, $\mathbf{F}$ a field of characteristic 0 or prime to the order of G, and let V and W be two $\mathbf{F}$-representations of G. Then*

(1) $i(V,\ W) = i(W,\ V)$;

(2) *if $V \cong \bigoplus_{\alpha \in A} p_\alpha M_\alpha$ and $W \cong \bigoplus_{\alpha \in A} q_\alpha M_\alpha$ are the decompositions of V and W into irreducibles, then*

$$i(V,\ W) = \sum_{\alpha \in A} p_\alpha q_\alpha \dim_{\mathbf{F}}(\operatorname{End}(M_\alpha));$$

(3) *if V is absolutely irreducible, then $i(V,\ W)$ is the multiplicity of V in W;*

(4) *V is absolutely irreducible if and only if $i(V,\ V) = 1$;*

(5) *if $\mathbf{F}$ is excellent for G and $\operatorname{char}(\mathbf{F}) = 0$, then*

$$i(V,\ W)\ =\ \langle V,\ W\rangle\ =\ \langle \chi_V,\ \chi_W\rangle\ =\ \frac{1}{n} \sum_{g \in G} \chi_V(g)\chi_W(g^{-1}).$$

Proof. Note that the hypotheses imply that R is semisimple, and recall Schur's lemma. The proof then becomes straightforward, and we leave it for the reader. $\qquad\square$

Remark. If G is finite, $\mathbf{F}$ is of characteristic zero or prime to the order of G, and V is of finite degree, then we have

$$i(V,\ W) = \dim_{\mathbf{F}} V^* \otimes_{\mathbf{F}(G)} W.$$

(See [AW, Theorem 7.1.28 and Proposition 7.2.32].)

Here we see the great utility of characters—if $\mathbf{F}$ is excellent for G and of characteristic zero, we may use them to compute intertwining numbers. In particular, we have:

Corollary 5.22.

(1) *Let $\mathbf{F}$ be an arbitrary field of characteristic 0. If W_1 and W_2 are two $\mathbf{F}$-representations of G of finite degree with $\chi_{W_1} = \chi_{W_2}$, then W_1 and W_2 are equivalent.*

(2) *Let $\mathbf{F}$ be an excellent field for G of characteristic $p \neq 0$. If W_1 and W_2 are two $\mathbf{F}$-representations of G of finite degree, having simple factorizations $W_i \cong \bigoplus m_\alpha^i M_\alpha$, $i = 1, 2$, with $\chi_{W_1} = \chi_{W_2}$, then $m_\alpha^1 \equiv m_\alpha^2 \pmod{p}$ for each α.*

(3) *Let $\mathbf{F}$ be an arbitrary field with $0 \neq \mathrm{char}(\mathbf{F})$ prime to n. If W_1 and W_2 are two absolutely irreducible $\mathbf{F}$-representations of G of finite degree with $\chi_{W_1} = \chi_{W_2}$, then W_1 and W_2 are equivalent.*

Proof.

(1) Let W_i have simple factorization $W_i \cong \bigoplus m_\alpha^i M_\alpha$, $i = 1, 2$. Then by Lemma 5.21 parts (2) and (5),

$$
\begin{aligned}
m_\alpha^1 &= \langle M_\alpha,\ W_1 \rangle / \dim_{\mathbf{F}} \mathrm{End}(M_\alpha) = \langle \chi_{M_\alpha}, \chi_{W_1} \rangle / \dim_{\mathbf{F}} \mathrm{End}(M_\alpha) \\
&= \langle \chi_{M_\alpha}, \chi_{W_2} \rangle / \dim_{\mathbf{F}} \mathrm{End}(M_\alpha) = \langle M_\alpha,\ W_2 \rangle / \dim_{\mathbf{F}} \mathrm{End}(M_\alpha) \\
&= m_\alpha^2
\end{aligned}
$$

for each α, so W_1 and W_2 are equivalent.

(2) In this case, $\dim_{\mathbf{F}} \mathrm{End}(M_\alpha) = 1$ for each α, so $m_\alpha^1 = m_\alpha^2$ in $\mathbf{F}$, i.e., $m_\alpha^1 = m_\alpha^2 \pmod{p}$.

(3) This follows directly from (2). $\qquad\square$

We could continue to work over a suitable excellent field of characteristic 0, but instead, for the sake of simplicity, we will take $\mathbf{F} = \mathbf{C}$ (the field of complex numbers). We note that $\mathbf{C}$ has the field automorphism $z \mapsto \bar{z}$ of complex conjugation, with $\overline{z_1 + z_2} = \bar{z}_1 + \bar{z}_2$, $\overline{z_1 z_2} = \bar{z}_1 \bar{z}_2$, and $\bar{\bar{z}} = z$.

If $\sigma : G \to \mathrm{Aut}(V)$ is any complex representation, its conjugate $\bar{\sigma} : G \to \mathrm{Aut}(V)$ is another representation, and $\chi_{\bar{\sigma}} = \overline{\chi}_\sigma$. By Lemma 5.21 (3), if G is finite, then σ is isomorphic to $\bar{\sigma}$ if and only if $\chi_\sigma = \chi_{\bar{\sigma}} = \overline{\chi}_\sigma$, i.e., if and only if χ_σ is real valued.

We have the following important result.

Lemma 5.23. *Let $\sigma : G \to \mathrm{Aut}(V)$ be a complex representation of the finite group G. Then for every $g \in G$,*

$$\chi_\sigma(g^{-1}) = \overline{\chi}_\sigma(g).$$

Proof. Let $g \in G$. Then g has finite order k, say. By Lemma 1.16, V has a basis $\mathcal{B}$ with $[\sigma(g)]_\mathcal{B}$ diagonal, and in fact,

$$[\sigma(g)]_\mathcal{B} = \mathrm{diag}(\zeta^{a_1}, \zeta^{a_2}, \ldots, \zeta^{a_d}),$$

where $\zeta = \exp(2\pi i/k)$, $d = \deg(V)$, and $a_1, \ldots, a_d \in \mathbb{Z}$. Then

$$[\sigma(g^{-1})]_\mathcal{B} = ([\sigma(g)]_\mathcal{B})^{-1} = \mathrm{diag}(\zeta^{-a_1}, \zeta^{-a_2}, \ldots, \zeta^{-a_d}).$$

But $\zeta^{-a} = \overline{\zeta^a}$ for any a, so

$$\chi_\sigma(g^{-1}) = \mathrm{Tr}\left([\sigma(g^{-1})]_\mathcal{B}\right) = \sum_{i=1}^{d} \zeta^{-a_i} = \sum_{i=1}^{d} \overline{\zeta^{a_i}} = \overline{\mathrm{Tr}\left([\sigma(g)]_\mathcal{B}\right)} = \overline{\chi_\sigma(g)}.$$

$\square$

Consequence 5.24. *If $\mathbf{F} = \mathbf{C}$, then $\chi_\sigma(g^{-1})$ may be replaced by $\overline{\chi}_\sigma(g)$ in results 5.8, 5.9, 5.12, 5.18, and 5.21.*

Lemma 5.23 also gives a handy way of encoding the orthogonality relations (Corollaries 5.8 and 5.18) for complex characters. Recall the character table $A = [a_{ij}] = [\chi_i(c_j)]$ of Definition 5.11. Let

$$C = [c_{ij}] = \left[\sqrt{\frac{|C_j|}{n}}\, a_{ij}\right] \in \mathrm{Mat}_t(\mathbf{C}).$$

Proposition 5.25. *The above matrix C is unitary, i.e., $C^{-1} = \overline{C}^{\,t}$.*

Proof. We leave this for the reader. $\square$

We now present some typical examples of the use of characters and their properties. In the first two examples we show how to use Frobenius's basic theorem, Theorem 3.3 (and the orthogonality relations, Corollary 5.8), to find all complex characters of a group (without first finding all the irreducible representations), and in the third we show how to use Lemma 5.21 to find the decomposition of a complex representation into irreducibles.

Example 5.26. Let us determine all the irreducible complex characters of the alternating group A_5 of order 60.

First we determine that A_5 has five conjugacy classes: $\{1\}$, $\{$products of two disjoint 2-cycles$\}$, $\{$3-cycles$\}$ each constitute a conjugacy class, while $\{$5-cycles$\}$ splits into two conjugacy classes (with the property that if g is in one of them, g^2 is in the other). Thus, as representatives for the conjugacy classes, we may take $c_1 = 1$, $c_2 = (1\,2)(3\,4)$, $c_3 = (1\,2\,3)$, $c_4 = (1\,2\,3\,4\,5)$, and $c_5 = (1\,3\,5\,2\,4)$. It may be checked that the conjugacy classes have sizes 1, 15, 20, 12, and 12, respectively. Thus there are five irreducible representations whose degrees d_i satisfy $\sum_{i=1}^{5} d_i^2 = 60$, which has the unique solution $d_1 = 1$, $d_2 = 3$, $d_3 = 3$, $d_4 = 4$, and $d_5 = 5$. Thus, we have determined the degrees of the irreducible representations.

Denote the irreducible representations by α_1, α_3, α_3', α_4, and α_5 with characters χ_1, χ_3, χ_3', χ_4, and χ_5. Of course, we have $\alpha_1 = \tau$, the trivial representation, with $\chi_1(g) = 1$ for all $g \in G$.

Consider the representation β on $\mathbf{C}^5$ given by letting A_5 act by permuting the coordinates, i.e.,

$$g([z_1, \ \ldots, \ z_5]) = [z_{g(1)}, \ \ldots, \ z_{g(5)}].$$

This is a permutation representation and the trace of any element is easy to compute. For $g \in A_5$, $\chi_\beta(g) = |\{i \mid g(i) = i\}|$. Of course $\chi_\beta(1) = 5$, and we observe that $\chi_\beta(c_2) = 1$, $\chi_\beta(c_3) = 2$, and $\chi_\beta(c_4) = \chi_\beta(c_5) = 0$. We compute $\langle \chi_\beta, \chi_\beta \rangle$ (noting that $\chi_\beta^* = \chi_\beta$) and obtain

$$\langle \chi_\beta, \chi_\beta \rangle = \frac{1}{60}(1 \cdot 5^2 + 15 \cdot 1^2 + 20 \cdot 2^2$$
$$+ \ 12 \cdot 0^2 + 12 \cdot 0^2)$$
$$= 2,$$

so β has two irreducible components. We compute

$$\langle \chi_1, \chi_\beta \rangle = \frac{1}{60}(1 \cdot 1 \cdot 5 + 15 \cdot 1 \cdot 1 + 20 \cdot 1 \cdot 2$$
$$+ \ 12 \cdot 1 \cdot 0 + 12 \cdot 1 \cdot 0)$$
$$= 1,$$

so $\tau = \alpha_1$ of degree one is one of them. The complement of τ is then an irreducible representation of degree 4, so is α_4. Furthermore, since $\beta = \tau \oplus \alpha_4$, we have $\chi_\beta = \chi_1 + \chi_4$, i.e., $\chi_4 = \chi_\beta - \chi_1$. We thus compute $\chi_4(1) = 4$, $\chi_4(c_2) = 0$, $\chi_4(c_3) = 1$, and $\chi_4(c_4) = \chi_4(c_5) = -1$.

Next we consider the representation γ on $\mathbf{C}^{10}$ given by the action of A_5 permuting coordinates, where now $\mathbf{C}^{10}$ is coordinatized by the ten sets $\{i, j\}$ of unordered pairs of distinct elements of $\{1, \ \ldots, \ 5\}$. Again,

$$\chi_\gamma(g) = |\{ \ \{i, j\} \mid g(\{i, j\}) = \{i, j\} \ \}|.$$

Then $\chi_\gamma(1) = 10$, $\chi_\gamma(c_2) = 2$, $\chi_\gamma(c_3) = 1$, and $\chi_\gamma(c_4) = \chi_\gamma(c_5) = 0$. Again, $\chi_\gamma^* = \chi_\gamma$, and

$$
\begin{aligned}
\langle \chi_\gamma, \chi_\gamma \rangle &= \frac{1}{60}(1 \cdot 10^2 + 15 \cdot 2^2 + 20 \cdot 1^2 \\
&\qquad + 12 \cdot 0^2 + 12 \cdot 0^2) \\
&= 3,
\end{aligned}
$$

so γ has three irreducible components. We compute

$$
\begin{aligned}
\langle \tau, \chi_\gamma \rangle &= \frac{1}{60}(1 \cdot 1 \cdot 10 + 15 \cdot 1 \cdot 2 + 20 \cdot 1 \cdot 1 \\
&\qquad + 12 \cdot 1 \cdot 0 + 12 \cdot 1 \cdot 0) \\
&= 1,
\end{aligned}
$$

and since we have already computed χ_4, we may compute

$$
\begin{aligned}
\langle \chi_4, \chi_\gamma \rangle &= \frac{1}{60}(1 \cdot 4 \cdot 10 + 15 \cdot 0 \cdot 2 + 20 \cdot 1 \cdot 1 \\
&\qquad + 12 \cdot (-1) \cdot 0 + 12 \cdot (-1) \cdot 0) \\
&= 1,
\end{aligned}
$$

so the complement of $\tau \oplus \alpha_4$ in γ is an irreducible representation of degree 5, so it is α_5. Also, $\chi_5 = \chi_\gamma - \chi_4 - \chi_1$, and we compute

$$
\chi_5(1) = 5, \qquad \chi_5(c_2) = 1, \qquad \chi_5(c_3) = -1, \qquad \chi_5(c_4) = \chi_5(c_5) = 0.
$$

We are left with the task of finding χ_3 and χ_3'. At this point, let us write down what we know of the character table of A_5.

	C_1	C_2	C_3	C_4	C_5
α_1	1	1	1	1	1
α_3	3	x_2	x_3	x_4	x_5
α_3'	3	y_2	y_3	y_4	y_5
α_4	4	0	1	-1	-1
α_5	5	1	-1	0	0

Our task is to determine the unknown entries. Set $z_i = x_i + y_i$ for $2 \le i \le 5$. First we note that if $\mathcal{R}$ is the regular representation,

$$
\mathcal{R} = \alpha_1 \oplus 3\alpha_3 \oplus 3\alpha_3' \oplus 4\alpha_4 \oplus 5\alpha_5,
$$

so

$$
\chi_\mathcal{R} = \chi_1 + 3\chi_3 + 3\chi_3' + 4\chi_4 + 5\chi_5.
$$

Evaluating this on c_2 gives (using Example 5.2 (3))

$$0 = 1 + 3z_2 + 4 \cdot 0 + 5 \cdot 1$$

so $z_2 = -2$, and, by evaluation on c_3, c_4, and c_5, we obtain $z_3 = 0$, $z_4 = z_5 = 1$. Now let us use orthogonality.

$$
\begin{aligned}
0 = \ \langle \chi_3, \chi_1 \rangle \ &= 1 \cdot 3 \cdot 1 + 15 \cdot x_2 \cdot 1 + 20 \cdot x_3 \cdot 1 \\
&\quad + 12 \cdot x_4 \cdot 1 + 12 \cdot x_5 \cdot 1, \\
0 = \ \langle \chi_3, \chi_4 \rangle \ &= 1 \cdot 3 \cdot 4 + 15 \cdot x_2 \cdot 0 + 20 \cdot x_3 \cdot 1 \\
&\quad + 12 \cdot x_4 \cdot (-1) + 12 \cdot x_5 \cdot (-1), \\
0 = \ \langle \chi_3, \chi_5 \rangle \ &= 1 \cdot 3 \cdot 5 + 15 \cdot x_2 \cdot 1 + 20 \cdot x_3 \cdot (-1) \\
&\quad + 12 \cdot x_4 \cdot 0 + 12 \cdot x_5 \cdot 0,
\end{aligned}
$$

so we obtain a linear system

$$
\begin{aligned}
15x_2 \ + \ 20x_3 \ + \ 12(x_4 + x_5) \ &= \ -3, \\
20x_3 \ - \ 12(x_4 + x_5) \ &= \ -12, \\
15x_2 \ - \ 20x_3 \ &= \ -15,
\end{aligned}
$$

with solution $x_2 = -1$, $x_3 = 0$, $x_4 + x_5 = 1$. Then also $y_2 = -1$, $y_3 = 0$, $y_4 + y_5 = 1$, and so, since $z_4 = 1$, $x_4 = y_5 = x$, say, $y_4 = x_5 = y$, say, with $x + y = 1$. Thus, it remains to determine x and y. At this point we may proceed in either of two ways:

(1) Note that c_4 is conjugate to c_4^{-1} and c_5 is conjugate to c_5^{-1}. We can thus apply Corollary 5.8 to conclude

$$1 = \ \langle \chi_3, \chi_3 \rangle \ = \ \frac{1}{60}(1 \cdot 3^2 + 15 \cdot (-1)^2 + 20 \cdot (0)^2 + 12x^2 + 12y^2)$$

so $x^2 + y^2 = 3$.

(2) Either x is real or it is not. If x is not real, then $\overline{\chi}_3 \neq \chi_3$ is also a character, so we must have $\chi_3' = \overline{\chi}_3$ and hence $y = \overline{x}$. Then

$$
\begin{aligned}
0 = \ \langle \chi_3, \chi_3' \rangle &= 1 \cdot (3)^2 + 14 \cdot (-1)^2 + 20 \cdot (0)^2 + 12x\overline{y} + 12y\overline{x} \\
&= 24 + 12(x^2 + \overline{x}^2),
\end{aligned}
$$

so $x^2 + \overline{x}^2 = -2$, which, together with $x + y = x + \overline{x} = 1$, gives $x = (1 + \sqrt{-5})/2$ and $y = (1 - \sqrt{-5})/2$ (or vice versa). But this is impossible, as the values of $\chi(c_4)$ must lie in the field $\mathbf{Q}(\sqrt[5]{1})$, and these numbers do not. Hence x and y are real, so $0 = \langle \chi_3, \chi_3' \rangle$ yields

$$0 = 24 + 12(xy + yx),$$

so $xy = -1$.

Now either pair of equations

$$x^2 + y^2 = 3,$$
$$x + y = 1$$

from method (1) or

$$xy = -1,$$
$$x + y = 1$$

from method (2) gives

$$x = (1 + \sqrt{5})/2 \quad \text{and} \quad y = (1 - \sqrt{5})/2$$

(or vice versa, but there is no order on χ_3 and χ_3', so we make this choice). Hence, the complete "expanded" character table of A_5 is

	C_1	C_2	C_3	C_4	C_5
	1	15	20	12	12
α_1	1	1	1	1	1
α_3	3	-1	0	$(1 + \sqrt{5})/2$	$(1 - \sqrt{5})/2$
α_3'	3	-1	0	$(1 - \sqrt{5})/2$	$(1 + \sqrt{5})/2$
α_4	4	0	1	-1	-1
α_5	5	1	-1	0	0

(We call this the "expanded" character table of A_5 because we have included on the second line, as is often but not always done, the number of elements in each conjugacy class. Note that there is *no* canonical order for either conjugacy classes or representations, so different character tables for the same group may "look" different.)

Example 5.27. Let us show how to determine all the irreducible complex characters of the symmetric group S_5 of order 120.

Two elements in a symmetric group are conjugate if and only if they have the same cycle structure, so S_5 has seven conjugacy classes, with representatives 1, (1 2), (1 2 3), (1 2)(3 4), (1 2 3 4), (1 2 3)(4 5), and (1 2 3 4 5), and we expect seven irreducible representations. One of them is τ, of course, and another, also of degree 1, is *sign* given by $sign(g) = $ the sign of the permutation $g = \pm 1 \in \text{Aut}(\mathbf{C})$.

Observe that the representations β and γ of A_5 constructed in Example 5.26 are actually restrictions of representations of S_5, so the representations α_4 and α_5 of A_5 constructed there are restrictions of representations $\widetilde{\alpha}_4$ and

$\widetilde{\alpha}_5$ of S_5. Since α_4 and α_5 are irreducible, so are $\widetilde{\alpha}_4$ and $\widetilde{\alpha}_5$. Furthermore, since $\widetilde{\alpha}_4$ and $\widetilde{\alpha}_5$ are irreducible, so are $sign \otimes \widetilde{\alpha}_4$ and $sign \otimes \widetilde{\alpha}_5$, and we may compute that the characters of $\widetilde{\alpha}_4$ and $sign \otimes \widetilde{\alpha}_4$ (respectively, $\widetilde{\alpha}_5$ and $sign \otimes \widetilde{\alpha}_5$) are unequal, so these representations are distinct.

Hence, we have found six irreducible representations, of degrees 1, 1, 4, 4, 5, 5, so we expect one more of degree d, with

$$1^2 + 1^2 + 4^2 + 4^2 + 5^2 + 5^2 + d^2 = 120,$$

so $d = 6$. If we call this $\widetilde{\alpha}_6$, then we have (using $\chi(\sigma)$ for χ_σ, for convenience),

$$\chi(\mathcal{R}) = \chi(\tau) + \chi(sign) + 4\chi(\widetilde{\alpha}_4) + 4\chi(sign \otimes \widetilde{\alpha}_4)$$
$$+ 5\chi(\widetilde{\alpha}_5) + 5\chi(sign \otimes \widetilde{\alpha}_5) + 6\chi(\widetilde{\alpha}_6)$$

enabling us to determine $\chi(\widetilde{\alpha}_6)$. We leave the details for the reader.

Example 5.28. Let $\zeta = \exp(2\pi i/3)$. It is easy to check from Example 4.3 that A_4 has the expanded character table (where we have listed the conjugacy classes in the same order as there).

	C_1	C_2	C_3	C_4
	1	3	4	4
$\tau = \pi^*(\theta_0)$	1	1	1	1
$\pi^*(\theta_1)$	1	1	ζ	ζ^2
$\pi^*(\theta_2)$	1	1	ζ^2	ζ
α	3	-1	0	0

We wish to find the decomposition of the tensor product of any two irreducible representations. The only nontrivial case is $\alpha \otimes \alpha$. Recall that $\chi_{\alpha \otimes \alpha} = (\chi_\alpha)(\chi_\alpha)$, and thus has values 9, 1, 0, 0 on the four conjugacy classes. We compute

$$\langle \alpha \otimes \alpha, \ \alpha \rangle = \frac{1}{12}(1 \cdot 3 \cdot 9 + 3 \cdot (-1) \cdot 1) = 2,$$

and for $i = 0$, 1, or 2

$$\langle \alpha \otimes \alpha, \ \pi^*(\theta_i) \rangle = \frac{1}{12}(1 \cdot 1 \cdot 9 + 3 \cdot 1 \cdot 1) = 1,$$

so

$$\alpha \otimes \alpha = 2\alpha \oplus \pi^*(\theta_0) \oplus \pi^*(\theta_1) \oplus \pi^*(\theta_2).$$

Remark 5.29. It is perhaps natural to conjecture that two groups with identical character tables must be isomorphic. This is, in fact, false! The groups D_8 and Q_8 (the dihedral and quaternion groups of order 8) are two distinct groups with the same character table. We leave the verification to the reader.

3.6. Idempotents and their Uses

In the last chapter we developed the theory of idempotents. In this section we relate them to characters, and we use them to prove several important results whose statements do not involve idempotents.

In the next two proofs, we regard characters as functions on $R = \mathbf{F}(G)$, extended from functions on G by linearity.

Proposition 6.1. *For $i = 1, \ldots, t$ let $e_{\alpha_i} \in \mathbf{F}(G)$ be defined by*

$$(6.1) \qquad e_{\alpha_i} = \frac{d_i}{n} \sum_{g \in G} \chi_i(g^{-1}) g,$$

where χ_i are the characters of the irreducible representations σ_i of G and $d_i = \deg(\sigma_i) = \chi_i(1)$. Then $\{e_{\alpha_i}\}$ is a complementary set of nonzero central-primitive idempotents in $R = \mathbf{F}(G)$.

Proof. Each e_{α_i} is nonzero as the coefficient of 1 in e_{α_i} is nonzero. We must show:

(1) $e_{\alpha_i}^2 = e_{\alpha_i}$ and $e_{\alpha_j} e_{\alpha_i} = 0$ for $i \neq j$,

(2) $e_{\alpha_1} + \cdots + e_{\alpha_t} = 1$,

(3) each e_{α_i} is central, and

(4) each e_{α_i} is central primitive.

(1) We compute

$$(6.2) \qquad e_{\alpha_j} e_{\alpha_i} = \left(\frac{d_j}{n}\right)\left(\frac{d_i}{n}\right) \sum_{g_1, g_2 \in G} \chi_j(g_2^{-1}) \chi_i(g_1^{-1}) g_2 g_1.$$

Setting $g_1 = g$, $g_2 = hg^{-1}$, and recalling that $\chi_j(gh^{-1}) = \chi_j(h^{-1}g)$, equation (6.2) becomes

$$e_{\alpha_j} e_{\alpha_i} = \left(\frac{d_j}{n}\right)\left(\frac{d_i}{n}\right) \sum_{h \in G} \left(\sum_{g \in G} \chi_j(h^{-1}g) \chi_i(g^{-1}) \right) h.$$

The interior sum, and hence the double sum, is zero if $i \neq j$ by Proposition 5.9. If $i = j$ the interior sum is $n\chi_i(h^{-1})/d_i$ (again by Proposition 5.9), and so in this case the double sum is

$$\frac{d_i^2}{n^2} \sum_{h \in G} \left(n\chi_i(h^{-1})/d_i \right) h = e_{\alpha_i}.$$

(2) Let χ be the character of the regular representation. Let

$$e_{\alpha_1} + \cdots + e_{\alpha_t} = \sum_{g \in G} a_g g,$$

so

$$h^{-1}(e_{\alpha_1} + \cdots + e_{\alpha_t}) = \sum_{g \in G} a_{hg} g.$$

Then, by Example 5.2 (3),

$$a_h = \frac{1}{n} \chi(h^{-1}(e_{\alpha_1} + \cdots + e_{\alpha_t})).$$

Now

$$\chi(h^{-1}(e_{\alpha_1} + \cdots + e_{\alpha_t})) = \sum_{j=1}^{t} \sum_{g \in G} \frac{d_j}{n} \chi_j(g^{-1}) \chi(h^{-1}g)$$

$$= \sum_{j=1}^{t} \sum_{g \in G} \frac{d_j}{n} \chi_j(g^{-1}) \sum_{i=1}^{t} d_i \chi_i(h^{-1}g)$$

$$= \sum_{j=1}^{t} \sum_{i=1}^{t} \frac{d_i d_j}{n} \sum_{g \in G} \chi_i(h^{-1}g) \chi_j(g^{-1})$$

$$= \sum_{i=1}^{n} \left(\frac{d_i^2}{n} \right) \left(n \chi_i(h^{-1})/d_i \right)$$

$$= \sum_{i=1}^{n} d_i \chi_i(h^{-1})$$

$$= \sum_{i=1}^{n} \chi_i(1) \chi_i(h^{-1})$$

$$= \begin{cases} 0 & \text{if } h \neq 1, \\ n & \text{if } h = 1, \end{cases}$$

where the fourth equality is by Proposition 5.9 and the last is by Corollary 5.18. Thus, $a_1 = 1$ and $a_h = 0$ for $h \neq 1$, giving $e_{\alpha_1} + \cdots + e_{\alpha_t} = 1$.

(3) It suffices to show that $e_{\alpha_i} h = h e_{\alpha_i}$, or equivalently, that $h^{-1} e_{\alpha_i} h = e_{\alpha_i}$, for each $h \in G$. But

$$h^{-1} e_{\alpha_i} h = \frac{d_i}{n} \sum_{g \in G} \chi_i(g^{-1}) h^{-1} g h = \frac{d_i}{n} \sum_{g \in G} \chi_i(h^{-1}g^{-1}h) h^{-1} g h$$

$$= \frac{d_i}{n} \sum_{g' \in G} \chi_i(g'^{-1}) g' = e_{\alpha_i},$$

since $g' = h^{-1}gh$ runs over G as g does.

(4) This follows directly from Lemma 2.3.19. $\qquad\square$

Proposition 6.2. *The representation of G on Re_{α_i} is isomorphic to $d_i\sigma_i$.*

Proof. Let ρ_i be the action of G on Re_{α_i}. We will show that $\chi(\rho_i) = d_i\sigma_i$.

Let ρ_0 be the regular representation with character χ_0. Then $\rho_0 = \bigoplus \rho_j$, so for every $r \in R$, $\chi_0(r) = \sum \chi(\rho_j(r))$. Set $r = ge_{\alpha_i}$. Then

$$\chi_0(ge_{\alpha_i}) = \sum \chi(\rho_j(ge_{\alpha_i})).$$

Now $\rho_j(ge_{\alpha_i}) = \rho_j(g)\rho_j(e_{\alpha_i})$. If $j \neq i$, then multiplication by e_{α_i} is zero on Re_j, so $\rho_j(ge_{\alpha_i}) = 0$. If $j = i$, then multiplication by e_{α_i} is the identity on Re_j, so $\rho_i(ge_{\alpha_i}) = \rho_i(g)$. Thus

$$\chi_0(ge_{\alpha_i}) = \chi(\rho_i(g)).$$

On the other hand, by Example 5.2 (3), the value of $\chi_0(ge_{\alpha_i})$ is n times the coefficient of g^{-1} in e_{α_i}, so by equation (6.1)

$$\chi_0(ge_{\alpha_i}) = d_i\chi_i(g).$$

Hence $\chi(\rho_i(g)) = d_i\chi_i(g) = \chi(d_i\sigma_i(g))$.

If $\mathrm{char}(\mathbf{F}) = 0$, this immediately implies that $\rho_i = d_i\sigma_i$. If $\mathrm{char}(\mathbf{F}) = p > 0$, we know that ρ_i is the direct sum of d_i copies of *some* representation, so this computation shows that representation must be σ_i (as $d_i \not\equiv 0 \pmod{p}$ by Corollary 5.7). $\qquad\square$

We will use idempotents to prove the following result:

Theorem 6.3. *Let $d_i = \deg(\sigma_i)$ for $i = 1, \ldots, t$ be the degrees of the irreducible complex representations of G. Then for each i, d_i divides n, the order of G.*

Proof. We begin with a general observation:

Claim. Let M be a finitely generated $\mathbb{Z}$-submodule of a $\mathbf{Q}$-vector space V. Suppose that $r \in \mathbf{Q}$ has the property that $rM \subseteq M$. Then $r \in \mathbb{Z}$.

Proof of claim. Since M is a finitely generated torsion-free $\mathbb{Z}$-module, M is free. Choose a basis for M and let m_1 be any element of that basis. Let $r = x/y$ for integers x and y. Then $rm_1 = m$ for some $m \in M$, so $xm_1 = ym$. But now, expressing m as a linear combination of elements of that basis, it is easy to see that y divides x.

We apply this here with $V = \mathbf{C}(G)$. Let $\zeta = \exp(2\pi i/u)$, with u the exponent of G. For fixed i, let M_i be the $\mathbb{Z}$-submodule of $\mathbf{C}(G)$ spanned by

$$\mathcal{B}_i = \{\zeta^k ge_{\alpha_i} \mid k = 0, \ldots, u - 1, \quad g \in G\}.$$

Since $e_{\alpha_i}^2 = e_{\alpha_i}$, we have by equation (6.1)

$$\frac{n}{d_i}e_{\alpha_i} = \frac{n}{d_i}e_{\alpha_i}^2 = \left(\frac{n}{d_i}e_{\alpha_i}\right)e_{\alpha_i} = \left(\sum_{g\in G}\chi_i(g^{-1})g\right)e_{\alpha_i} \in M_i,$$

as it is a $\mathbb{Z}$-linear combination of elements of $\mathcal{B}_i$, and similarly for the other elements of $\mathcal{B}_i$. This immediately implies $(n/d_i)(M_i) \subseteq M_i$, so $n/d_i \in \mathbb{Z}$ and d_i divides n, as claimed. $\square$

Notation. Let $\mathbf{F}_0 \subset \mathbf{F}$ be the prime field. In other words, if $\mathrm{char}(\mathbf{F}) = 0$, then $\mathbf{F}_0 = \mathbf{Q}$, while if $\mathrm{char}(\mathbf{F}) = p > 0$, $\mathbf{F}_0 = \mathbf{F}_p$.

For σ an $\mathbf{F}$-representation of G with character $\chi = \chi_\sigma$, let $\mathbf{F}_0(\chi)$ be the extension of $\mathbf{F}_0$ generated by the values $\{\chi(g)\}_{g\in G}$ of the character χ on the elements of G. Certainly if σ is defined over $\mathbf{F}$, then $\mathbf{F} \supseteq \mathbf{F}_0(\chi_\sigma)$. The converse is not in general true, but we can conclude the following:

Corollary 6.4. *Let σ be an absolutely irreducible representation of G over $\mathbf{F}$, with character χ_σ. If $d = \deg(\sigma)$, then $d\sigma$ is defined over $\mathbf{F}_0(\chi_\sigma)$.*

Proof. For each i, the idempotent e_{α_i} is evidently defined over $\mathbf{F}_0(\chi_{\sigma_i})$, and then the result is immediate from Proposition 6.2. $\square$

Remark. As an example, let ρ be the irreducible 2-dimensional complex representation of $\mathbf{Q}_8$, given in Example 4.2. Then $\mathbf{F}_0(\chi_\rho) = \mathbf{R}$, but ρ is *not* defined over $\mathbf{R}$, while 2ρ *is*. (See Example 7.12.)

Remark. Let $\mathrm{char}(\mathbf{F}) = 0$. From Proposition 2.4.17 we know that there is an integer s such that $k\sigma$ is defined over $\mathbf{F}_0(\chi_\sigma)$ if and only if s divides k. This integer s is called the **Schur index** of σ. From Corollary 6.4 we see that for any absolutely irreducible representation σ of G in characteristic 0, the Schur index of σ divides the degree of σ.

Of course, we can consider the analogous question in positive characteristic. Here it turns out that every irreducible representation σ *is* defined over $\mathbf{F}_0(\chi_\sigma)$. We have:

Theorem 6.5. *Let $\mathbf{F}$ be a finite field of characteristic prime to n, and let $\mathbf{E}$ be a subfield of $\overline{\mathbf{F}}$.*

(1) *If σ is an irreducible $\overline{\mathbf{F}}$-representation of G with $\chi_\sigma(g) \in \mathbf{E}$ for every $g \in G$, then σ is defined over $\mathbf{E}$.*

(2) *Every $\overline{\mathbf{F}}$-representation of G is defined over the field generated by adjoining*

$$\{\chi_\sigma(g) \mid \sigma \text{ an irreducible } \overline{\mathbf{F}}\text{-representation of } G,\ g \in G\}$$

to $\mathbf{F}$.

(3) *Every $\overline{\mathbf{F}}$-representation of G is defined over the field $\mathbf{F}(\sqrt[u]{1})$.*

Proof. (1) If e_α is the central-primitive idempotent corresponding to the irreducible representation σ, then $\chi_\sigma(g) \in \mathbf{E}$ for all $g \in G$ gives $e_\sigma \in \mathbf{E}(G)$ by Propositions 6.1 and 6.2. Then the result follows immediately from Proposition 2.4.18.

(2) This follows immediately from (1).

(3) Since $\chi_\sigma(g) \in \mathbf{F}(\sqrt[u]{1})$ for every σ and every $g \in G$, this is a special case of (2). $\qquad\square$

Next we shall see that every absolutely irreducible representation of G in positive characteristic "comes from" an absolutely irreducible representation of G in characteristic 0 of the same degree (providing we remain in the semisimple case). Thus the degrees of the absolutely irreducible representations of G are independent of the characteristic (providing that the characteristic p is relatively prime to n).

We identify the ring $\mathbb{Z}(\sqrt[u]{1})$ with the ring $\mathbb{Z}[X]/(X^u - 1)$, and the field $\mathbf{F}_p(\sqrt[u]{1})$ with the ring $(\mathbb{Z}/p\mathbb{Z})[X]/(X^u - 1)$. Then the map $\mathbb{Z} \to \mathbb{Z}/p\mathbb{Z}$ given by reduction mod p extends to a map $\varphi : \mathbb{Z}(\sqrt[u]{1}) \to \mathbf{F}_p(\sqrt[u]{1})$. This map further extends to $\varphi : \mathbb{Z}'(\sqrt[u]{1}) \to \mathbf{F}_p(\sqrt[u]{1})$, where $\mathbb{Z}' \subset \mathbf{Q}$ is the subring of rational numbers whose denominators are relatively prime to p (by $\varphi(a/b) = \varphi(a)\varphi(b)^{-1}$), and hence to a map

$$\varphi : \mathbb{Z}'(\sqrt[u]{1})(G) \to \mathbf{F}_p(\sqrt[u]{1})(G).$$

Lemma 6.6. *Let $\overline{e}_{\alpha_i} = \varphi(e_{\alpha_i}), i = 1,\ldots,t$. Then $\{\overline{e}_{\alpha_1},\ldots,\overline{e}_{\alpha_t}\}$ is a complementary set of central-primitive idempotents.*

Proof. First note that, by (6.1) and the assumption that p does not divide n, $e_{\alpha_i} \in \mathbb{Z}'(\sqrt[u]{1})(G)$, so $\varphi(e_{\alpha_i})$ is defined, for each i. Second, note that, by (6.1), the coefficient of the group element 1 in e_{α_i} is d_i^2/n, and p does not divide d_i by Theorem 6.3, so $\overline{e}_{\alpha_i} = \varphi(e_{\alpha_i})$ is nonzero, for each i. Since φ is a homomorphism, and $\{e_{\alpha_1},\ldots,e_{\alpha_t}\}$ is a complementary set of central idempotents, so is $\{\overline{e}_{\alpha_1},\ldots,\overline{e}_{\alpha_t}\}$. But $\dim_{\mathbf{F}} \text{Center}(\mathbf{F}(G)) = t$ by Lemma 3.2, so each $\overline{e}_{\alpha_i}$ is central primitive by Lemma 2.3.19. $\qquad\square$

Thus the block decomposition

$$\overline{\mathbf{Q}}(G) = \bigoplus \overline{\mathbf{Q}}(G)e_{\alpha_i}$$

gives a block decomposition

$$\overline{\mathbf{F}}_p(G) = \bigoplus \overline{\mathbf{F}}_p(G)\overline{e}_{\alpha_i}.$$

The dimension of each block is the square of the dimension of the corresponding irreducible representation. We have denoted the dimensions of these representations by $\{d_i\}$ for $\overline{\mathbf{Q}}$; let us denote them by $\{\overline{d}_i\}$ for $\overline{\mathbf{F}}_p$.

Theorem 6.7. *For each* i, $\overline{d}_i = d_i$.

Proof. Each i, let $\{\overline{v}_j\}$, $j = 1, \dots, \overline{d}_i^{\,2}$, be a basis of $\mathbf{F}_p(\sqrt[u]{1})(G)\overline{e}_{\alpha_i}$. Since φ is an epimorphism, we may find v_j with $\varphi(v_j) = \overline{v}_j$. Since $\{\overline{v}_j\}$ is linearly independent, so is $\{v_j\}$. (We prove the contrapositive. Suppose that $\{v_j\}$ is linearly dependent, so that there is a nontrivial linear combination $\sum c_j v_j = 0$ with $c_j \in \mathbb{Z}'(\sqrt[u]{1})(G)$. After dividing by the appropriate power of p (which may be $p^0 = 1$), we may assume some $c_{j_0} \in \mathbb{Z}'(\sqrt[u]{1})(G)$ but $c_{j_0} \notin p\mathbb{Z}'(\sqrt[u]{1})(G)$. Then we have $0 = \varphi(0) = \varphi(\sum c_j v_j) = \sum \varphi(c_j)\overline{v}_j$ with $\varphi(c_{j_0}) \neq 0$, so $\{\overline{v}_j\}$ is linearly dependent.)

But $\{v_j\}$ is a set of elements in $\mathbf{Q}(\sqrt[u]{1})(G)e_{\alpha_i}$, of dimension d_i^2, so $\overline{d}_i^{\,2} \leq d_i^2$. But $n = \sum_{i=1}^t \overline{d}_i^{\,2} \leq \sum_{i=1}^t d_i^2 = n$, so $\overline{d}_i = d_i$ for each i, as claimed. $\square$

3.7. Subfields of the Complex Numbers

In this section we investigate representations that are defined over, or whose characters are in, subfields $\mathbf{F}$ of $\mathbf{C}$, with particular attention to the cases $\mathbf{F} = \mathbf{Q}$ and $\mathbf{F} = \mathbf{R}$.

Theorem 7.1. *Let G be a finite group.*

(1) *Let $g \in G$ have the property that for every m relatively prime to the order of g, g and g^m are conjugate. Then for every complex representation σ of G, $\chi(g)$ is rational (where $\chi = \chi_\sigma$).*

(2) *Let σ be a complex representation of G such that $\chi = \chi_\sigma$ is $\mathbf{Q}$-valued. Then for any $g \in G$, and any m relatively prime to the order of g, $\chi(g) = \chi(g^m)$.*

(3) *The following are equivalent:*

    (i) *For every complex representation σ of G, and every $g \in G$, $\chi(g)$ is rational (where $\chi = \chi_\sigma$).*

    (ii) *For every $g \in G$, and every m relatively prime to the order of g, g and g^m are conjugate.*

Proof. Let g have order k, and fix a k-th root of unity ζ_k. Note that instead of considering complex representations, we may without loss of generality consider representations defined over $\overline{\mathbf{Q}}$. Also, if σ is a representation of G

defined over $\overline{\mathbf{Q}}$, and $\gamma \in \mathrm{Gal}(\overline{\mathbf{Q}}/\mathbf{Q})$, then σ^γ is also a representation of G over $\overline{\mathbf{Q}}$. If σ has character χ, we let χ^γ denote the character of σ^γ.

(1) For each $\gamma \in \mathrm{Gal}(\overline{\mathbf{Q}}/\mathbf{Q})$, $\gamma(\zeta_k) = \zeta_k^m$ for some m with $(m,k) = 1$. Then $\chi^\gamma(g) = \chi(g^m)$. But g and g^m are conjugate, so $\chi(g^m) = \chi(g)$, and hence $\chi^\gamma(g) = \chi(g)$. Thus $\chi(g)$ is fixed by every $\gamma \in \mathrm{Gal}(\overline{\mathbf{Q}}/\mathbf{Q})$, so $\chi(g) \in \mathbf{Q}$.

(2) For any m with $(m,k) = 1$, there is a $\gamma \in \mathrm{Gal}(\overline{\mathbf{Q}}/\mathbf{Q})$ such that $\gamma(\zeta_k) = \zeta_k^m$. Then $\chi^\gamma(g) = \chi(g^m)$. But $\chi(g) \in \mathbf{Q}$, so $\chi^\gamma(g) = \chi(g)$, and hence $\chi(g) = \chi(g^m)$.

(3) (ii) $\Rightarrow$ (i) The hypothesis of (1) is satisfied for every $g \in G$.

(3) (i) $\Rightarrow$ (ii) By (2), $\chi(g) = \chi(g^m)$ for every complex character χ of G, so g and g^m are conjugate by Proposition 5.16. $\qquad\square$

Remark. As an important application of this theorem, note that (3) (ii), and hence (3) (i), is true for every symmetric group. In fact, all representations of every symmetric group are defined over $\mathbf{Q}$. It also applies to D_8, which has all of its complex representations defined over $\mathbf{Q}$, and to Q_8, which does not. (See Example 7.12.)

Let $\mathbf{F} \subseteq \mathbf{C}$ and set $\mathbf{E} = \mathbf{F}(\sqrt[u]{1})$. Then $\mathrm{Gal}(\mathbf{E}/\mathbf{F})$ is isomorphic to a subgroup H of $\mathrm{Gal}(\mathbf{Q}(\sqrt[u]{1})/\mathbf{Q})$. Consider an element γ of this group. Then, setting $\zeta = \exp(2\pi i/u)$, the action of γ is given by $\gamma(\zeta) = \zeta^\kappa$ for some integer κ prime to u. Moreover, as γ ranges over H, κ ranges over a subgroup K of the residue classes of integers modulo u, under multiplication.

Call two elements g_1 and g_2 of G K-**conjugate** if g_1 is conjugate to g_2^κ for some $\kappa \in K$. The fact that K is a group readily implies that K-conjugacy is an equivalence relation. (Of course, if g_1 and g_2 are conjugate, they are also K-conjugate.) While H and K depend on $\mathbf{E}$ and $\mathbf{F}$, we suppress this in the notation.

Proposition 7.2. *Let $\mathbf{F} \subseteq \mathbf{C}$ be a subfield. The number of irreducible $\mathbf{F}$-representations of G is equal to the number of K-conjugacy classes of elements of G.*

Proof. Let $\rho_1, \ldots, \rho_s$ be the irreducible $\mathbf{F}$-representations of G with characters $\widetilde{\chi}_1, \ldots, \widetilde{\chi}_s$. First observe that each $\widetilde{\chi}_j$ is constant on K-conjugacy classes, as for $\gamma \in H$, $\widetilde{\chi}_j(g) = \gamma(\widetilde{\chi}_j(g)) = \widetilde{\chi}_j(g^\kappa)$ for some $\kappa \in K$, and as γ runs over H, κ runs over K. Furthermore, $\{\widetilde{\chi}_1, \ldots, \widetilde{\chi}_s\}$ are certainly linearly independent. We will show that they span $\{\, f : G \to \mathbf{C} \mid f$ is constant on K-conjugacy classes$\}$. This shows that they are a basis for this space. Hence s is equal to the dimension of this space, which is equal to the number of K-conjugacy classes in G.

Let f be an arbitrary function that is constant on K-conjugacy classes. Then f is constant on conjugacy classes, i.e., f is a class function, so

$f = \sum_{i=1}^{t} c_i \chi_i$ for some $c_i \in \mathbf{C}$. Then $f(g) = f(g^\kappa) = \sum c_i \chi_i(g^\kappa)$, so $\sum_{\kappa \in K} f(g^\kappa) = kf(g)$, where k is the order of K. Thus

$$f(g) = (1/k) \sum_{\kappa \in K} \sum_i c_i \chi_i(g^\kappa)$$

$$= (1/k) \sum_{\gamma \in H} \sum_i c_i \gamma(\chi_i(g))$$

$$= (1/k) \sum_i c_i \sum_{\gamma \in H} \gamma(\chi_i(g)).$$

But $\sum_{\gamma \in H} \gamma(\chi_i(g))$ is a $\mathbf{F}$-valued character, as it is invariant under $H = \mathrm{Gal}(\mathbf{F}(\sqrt[u]{1})/\mathbf{F}))$, so by Corollary 6.4 some multiple m_i of it is the character of an $\mathbf{F}$-representation, i.e., for some integers $\{n_{j,i}\}$,

$$m_i \sum_{\gamma \in H} \gamma(\chi_i(g)) = \sum_{j=1}^{s} n_{j,i} \widetilde{\chi}_j(g).$$

Assembling all this, we find

$$f = \sum_j \left(\sum_i \frac{c_i n_{j,i}}{k m_i} \right) \widetilde{\chi}_j$$

as claimed. $\square$

Corollary 7.3.

(1) *The number of irreducible $\mathbf{Q}$-representations of G is equal to the number of conjugacy classes of cyclic subgroups of G.*

(2) *The number of irreducible $\mathbf{R}$-representations of G is equal to the number of equivalence classes of elements of G under the relation $g_1 \equiv g_2$ if g_1 is conjugate to $g_2^{\pm 1}$.*

Proof.

(1) Here $\mathbf{F} = \mathbf{Q}$, $\mathbf{E} = \mathbf{Q}(\sqrt[u]{1})$, and so two elements of G are K-conjugate if and only if the cyclic subgroups they generate are conjugate.

(2) Here $\mathbf{F} = \mathbf{R}$. If $u = 2$, then $\mathbf{E} = \mathbf{R}$, G is abelian, and the result is trivial. If $u > 2$, then $\mathbf{E} = \mathbf{C}$, $\mathrm{Gal}(\mathbf{E}/\mathbf{F}) \cong \mathbb{Z}/2\mathbb{Z}$, and the nontrivial element acts by complex conjugation, $\zeta \mapsto \overline{\zeta} = \zeta^{-1}$, so $K = \{\pm 1\}$. $\square$

Remark. We emphasize that Proposition 7.2 is concerned with irreducible $\mathbf{F}$-representations of G, *not* irreducible complex representations of G that are defined over $\mathbf{F}$. (Certainly an irreducible complex representation of G that is defined over $\mathbf{F}$ is an irreducible $\mathbf{F}$-representation of G, but an irreducible

F-representation of G may not be irreducible as a complex representation of G.)

Now we turn to real characters.

Proposition 7.4. *Let G be a finite group. Then every complex character of G is real valued if and only if every element in G is conjugate to its own inverse.*

Proof. Suppose that every g is conjugate to g^{-1}. Then for every complex character χ, $\chi(g) = \chi(g^{-1}) = \overline{\chi(g)}$, so $\chi(g)$ is real.

Conversely, suppose that $\chi_i(g)$ is real for every irreducible complex representation σ_i and every $g \in G$. Since $\sigma_1 = \tau$, $\chi_1(g) = 1$, so

$$0 < \sum_{j=1}^{t} \chi_j(g)^2 = \sum_{j=1}^{t} \chi_j(g)\chi_j((g^{-1})^{-1})$$

so g and g^{-1} are conjugate by Corollary 5.18. $\qquad\square$

Let us call a conjugacy class C_i **self-inversive** if for some (and hence for every) $c_i \in C_i$, we also have $c_i^{-1} \in C_i$.

Proposition 7.5. *The number of self-inversive conjugacy classes of G is equal to the number of irreducible complex characters of G that are real valued.*

Proof. We outline the proof and leave the details to the reader. We make use of the matrix C of Proposition 5.25. Consider the diagonal elements of CC^t. From the fact that $\overline{\chi}_i = \chi_i$ if χ_i is real valued and is the character of a distinct irreducible representation if not, we see that the i-th diagonal entry of CC^t is 1 if χ_i is real valued and 0 if not. Thus $\mathrm{Tr}(CC^t)$ is equal to the number of real-valued characters.

Now consider the diagonal elements of C^tC. From the fact that $c_i^{-1} \in C_i$ if C_i is self-inversive, and is in a different conjugacy class if not, we see that the i-th diagonal entry of C^tC is 1 if C_i is self-inversive and 0 if not. Thus $\mathrm{Tr}(C^tC)$ is equal to the number of self-inversive conjugacy classes, and since $\mathrm{Tr}(C^tC) = \mathrm{Tr}(CC^t)$, the proposition follows. $\qquad\square$

Corollary 7.6. *If G has odd order, no nontrivial irreducible complex character of G is real valued.*

Proof. Suppose that $ghg^{-1} = h^{-1}$. Then $g^i h g^{-i} = h$ if i is even or h^{-1} if i is odd. In particular, $g^n h g^{-n} = h^{-1}$ (recalling that $n = |G|$, which is odd). But $g^n = 1$, so this gives $h = h^{-1}$, i.e., $h^2 = 1$. Since n is odd, this gives $h = 1$, and the corollary follows from Proposition 7.5. $\qquad\square$

Let us consider a single irreducible complex representation σ of G with character χ. By Proposition 2.4.22, $\sigma \oplus \overline{\sigma}$ is defined over $\mathbf{R}$ (as complex conjugation is the nontrivial element of $\mathrm{Gal}(\mathbf{C}/\mathbf{R})$). Of course, the representation $\overline{\sigma}$ has character $\overline{\chi}$. Clearly, if χ is not real valued, $\overline{\chi} \neq \chi$, and σ is not defined over $\mathbf{R}$. But if χ is real valued, then $\overline{\chi} = \chi$, so $\overline{\sigma}$ is equivalent to σ, and then 2σ is defined over $\mathbf{R}$. This still leaves the question of whether σ is defined over $\mathbf{R}$. A character computation provides the answer.

Definition 7.7. Let σ be a complex representation of G, with character χ. The **Frobenius-Schur indicator** $\mathrm{fs}(\sigma)$ is

$$\mathrm{fs}(\sigma) = \frac{1}{n} \sum_{g \in G} \chi(g^2).$$

Proposition 7.8. *Let σ be an irreducible complex representation of G, $\sigma : G \to \mathrm{Aut}(M)$, with character χ. Then*

(1) $\mathrm{fs}(\sigma) = 0$ if and only if χ is not real valued.

(2) $\mathrm{fs}(\sigma) = 1$ if and only if χ is real valued and M admits a nonsingular G-invariant symmetric bilinear form.

(3) $\mathrm{fs}(\sigma) = -1$ if and only if χ is real valued and M admits a nonsingular G-invariant skew-symmetric bilinear form.

Proof. Observe that $\langle \sigma, \overline{\sigma} \rangle = \frac{1}{n} \sum_{g \in G} \chi(g)^2 = \langle \sigma \otimes \sigma, \tau \rangle$. Thus if $\chi \neq \overline{\chi}$, so σ is not isomorphic to $\overline{\sigma}$, τ does not appear in $\sigma \otimes \sigma$, while if χ is real valued, so $\chi = \overline{\chi}$, then τ appears in $\sigma \otimes \sigma$ with multiplicity 1.

Let $\sigma : G \to \mathrm{Aut}(M)$, so $\sigma \otimes \sigma : G \to \mathrm{Aut}(M \otimes M)$. Set $T(M) = M \otimes M$. Let $\varphi : T(M) \to T(M)$ be defined by $\varphi(m_1 \otimes m_2) = m_2 \otimes m_1$. Note that φ is a map of representations. Then $\varphi^2 = 1$ so $T(M)$ splits into the $+1$ and -1 eigenspaces of φ as representations of G, $T(M) = S(M) \oplus E(M)$. Let χ_T, χ_S and χ_E be the characters of the representations $\sigma_T, \sigma_S, \sigma_E$ of G on $T(M), S(M)$, and $E(M)$, respectively.

Thus we see that if χ is not real valued, τ does not appear in $T(M)$, while if χ is real valued, either τ appears in $S(M)$ with multiplicity 1 and not in $E(M)$, or vice versa.

Let us compute χ_T, χ_S, and χ_E. For $g \in G$, pick a basis of M in which $\sigma(g)$ is diagonal. Let that basis be $\{v_1, \ldots, v_d\}$, and let the diagonal entries of $\sigma(g)$ be $\lambda_1, \ldots, \lambda_d$. Then $T(M)$ has basis $\{v_i \otimes v_j\}$, and in that basis $\sigma_T(g)$ has diagonal entries $\{\lambda_i \lambda_j\}$, so $\chi_T(g) = \sum_{i,j} \lambda_i \lambda_j = (\sum_i \lambda_i)(\sum_j \lambda_j) = \chi(g)^2$. (Of course, we already know this.) But then also $S(M)$ has basis $\{v_i \otimes v_j + v_j \otimes v_i \mid i \leq j\}$ and $E(M)$ has basis $\{v_i \otimes v_j - v_j \otimes v_i \mid i < j\}$, so in those bases $\sigma_S(g)$ has diagonal entries $\{\lambda_i \lambda_j \mid i \leq j\}$ and $\sigma_E(g)$ has diagonal entries $\{\lambda_i \lambda_j \mid i < j\}$, so $\chi_S(g) = \sum_{i \leq j} \lambda_i \lambda_j$ and $\chi_E(g) = \sum_{i < j} \lambda_i \lambda_j$.

Of course $\sigma(g^2)$ has diagonal entries (in that basis of M) $\lambda_1^2, \ldots, \lambda_d^2$. We have the algebraic identity

$$\sum \lambda_i^2 = \left(\sum \lambda_i\right)^2 - 2 \sum_{i<j} \lambda_i \lambda_j,$$

so

$$\chi(g^2) = \chi_T(g) - 2\chi_E(g).$$

Summing over $g \in G$, and dividing by n, gives

$$\begin{aligned}
\mathrm{fs}(\chi) &= \frac{1}{n} \sum_{g \in G} \chi_T(g) - 2 \cdot \frac{1}{n} \sum_{g \in G} \chi_E(g) \\
&= \frac{1}{n} \sum_{g \in G} \chi_S(g) - \frac{1}{n} \sum_{g \in G} \chi_E(g) \\
&= \langle \chi_S, \tau \rangle - \langle \chi_E, \tau \rangle
\end{aligned}$$

(where the second equality is because $\chi_T(g) = \chi_S(g) + \chi_E(g)$).

As we have seen, if χ is not real valued, τ does not appear in $T(M)$, hence in neither $S(M)$ nor $E(M)$, so $\mathrm{fs}(\chi) = 0$, and conversely. If χ is real valued, then τ appears in exactly one of χ_S and χ_E, so $\mathrm{fs}(\chi) = 1$ if τ appears in χ_S and $\mathrm{fs}(\chi) = -1$ if τ appears in χ_E, and conversely.

Suppose τ appears in χ_S. Then there is a nontrivial map of representations $f : S(M) \to \mathbf{C}$. This defines a symmetric bilinear form on M by

$$\langle m_1, m_2 \rangle = f(m_1 \otimes m_2 + m_2 \otimes m_1),$$

and this form is nonsingular as $\{m \mid f(m, m') = 0 \text{ for all } m' \in M\}$ would be a subrepresentation of M, but M is irreducible.

Similarly, if τ appears in $E(M)$, there is a nontrivial map of representations $f : E(M) \to \mathbf{C}$, which defines a nonsingular skew-symmetric bilinear form f on M by

$$\langle m_1, m_2 \rangle = f(m_1 \otimes m_2 - m_2 \otimes m_1).$$

$\square$

Proposition 7.9. *Let σ be an irreducible complex representation of G, $\sigma : G \to \mathrm{Aut}(M)$. The following are equivalent:*

(1) *M is defined over $\mathbf{R}$.*

(2) *M admits a nonsingular G-invariant symmetric bilinear form.*

Proof. Suppose that M is defined over $\mathbf{R}$, so $M = \mathbf{C} \otimes_{\mathbf{R}} M_0$ for some $\mathbf{R}(G)$-module M_0. Choose any positive-definite nonsingular bilinear form $\langle\,,\,\rangle'$ on M_0. (For example, pick a basis of M_0 and choose the usual dot product.) Then average $\langle\,,\,\rangle'$ over G to get a nonsingular G-invariant symmetric bilinear form $\langle\,,\,\rangle_0$ defined by

$$\langle m_1, m_2 \rangle_0 \;=\; \frac{1}{n} \sum_{g \in G} \langle g m_1, g m_2 \rangle'.$$

(Note $\langle\,,\,\rangle_0$ is nonsingular as $\langle m, m \rangle$ is positive for every $m \neq 0$.) This gives a nonsingular bilinear form $\langle\,,\,\rangle$ on M defined by

$$\langle c_1 \otimes m_1, c_2 \otimes m_2 \rangle \;=\; c_1 c_2 \langle m_1, m_2 \rangle_0.$$

Now we prove the converse. First we make a general observation. We know that $M \oplus \overline{M}$ is defined over $\mathbf{R}$, so $M \oplus \overline{M} \cong \mathbf{C} \otimes_{\mathbf{R}} M_0$ for some M_0. Suppose M_0 is reducible, $M_0 = N_1 \oplus N_2$. Then

$$M \oplus \overline{M} \cong \mathbf{C} \otimes_{\mathbf{R}} M_0 \cong \mathbf{C} \otimes_{\mathbf{R}} (N_1 \oplus N_2) \cong (\mathbf{C} \otimes_{\mathbf{R}} N_1) \oplus (\mathbf{C} \otimes_{\mathbf{R}} N_2)$$

so, since M (and hence $\overline{M}$) is irreducible, $M \cong \mathbf{C} \otimes_{\mathbf{R}} N_1$ and $\overline{M} \cong \mathbf{C} \otimes_{\mathbf{R}} N_2$, or vice versa, but, in any case, this means that M is defined over $\mathbf{R}$. Thus we need only show that if M admits a nontrivial G-invariant symmetric bilinear form, then M_0 is reducible (where M_0 is such that $M \oplus \overline{M} = \mathbf{C} \otimes_{\mathbf{R}} M_0$).

Let M have basis $\{m_j\}$, $\sigma : G \to \mathrm{Aut}(M)$, and give $\overline{M}$ the basis $\{\overline{m}_j\}$, so $\overline{\sigma} : G \to \mathrm{Aut}(M)$ is given by $[\overline{\sigma}(g)] = \overline{[\sigma(g)]}$, where $[\ ]$ denotes matrix in the respective basis. Set
$$n_j = m_j + i\overline{m}_j,$$
$$n_j' = i m_j + \overline{m}_j,$$

and let M_0 be the *real* span of $\{n_j, n_j'\}$. Note this set is linearly independent over $\mathbf{C}$, so the *complex* span of $\{n_j, n_j'\}$ is $M \oplus \overline{M}$ and hence $M \oplus \overline{M} = \mathbf{C} \otimes_{\mathbf{R}} M_0$, as vector spaces.

Direct computation shows that M_0 is invariant under the action of G, so M_0 is a real representation of G and $M \oplus \overline{M} = \mathbf{C} \otimes_{\mathbf{R}} M_0$ as representations of G.

Give M the nonsingular G-invariant symmetric bilinear form $\langle\,,\,\rangle$ defined above, and give $\overline{M}$ the nonsingular G-invariant symmetric bilinear form $\langle\,,\,\rangle^-$, where $\langle \overline{m}, \overline{m}' \rangle^- = \overline{\langle m, m' \rangle}$ for $\overline{m}, \overline{m}' \in \overline{M}$.

Give $M \oplus \overline{M}$ the G-invariant symmetric bilinear form

$$\langle\!\langle\,,\,\rangle\!\rangle \;=\; \frac{1}{2i} (\langle\,,\,\rangle \oplus \langle\,,\,\rangle^-),$$

i.e., the form defined by

$$\langle\langle m, m'\rangle\rangle = \frac{1}{2i}\langle m, m'\rangle \quad \text{for } m, m' \in M$$

$$\langle\langle \overline{m}, \overline{m}'\rangle\rangle = \frac{1}{2i}\langle \overline{m}, \overline{m}'\rangle^{-} \quad \text{for } \overline{m}, \overline{m}' \in \overline{M}$$

$$\langle\langle m, \overline{m}'\rangle\rangle = 0 \quad \text{for } m \in M, \ \overline{m} \in \overline{M}.$$

Let $\langle\langle\ ,\ \rangle\rangle_0$ be the restriction of $\langle\langle\ ,\ \rangle\rangle$ to $M_0 \otimes_{\mathbf{R}} M_0$. Direct computation shows that $\langle\langle\ ,\ \rangle\rangle_0$ is a real-valued (due to the factor $1/2i$) nonsingular (as $\langle\ ,\ \rangle$ is nonsingular) symmetric bilinear form on M_0. Indeed,

$$(7.1) \qquad \langle\langle n_j, n_j\rangle\rangle_0 = \frac{1}{2i}(\langle m_j, m_j\rangle - \overline{\langle m_j, m_j\rangle})$$

and

$$(7.2) \qquad \langle\langle n'_j, n'_j\rangle\rangle_0 = \frac{1}{2i}(\overline{\langle m_j, m_j\rangle} - \langle m_j, m_j\rangle).$$

Now by the general theory of bilinear forms, M_0 has a direct sum decomposition $M_0 = M_0^{+} \oplus M_0^{-}$, where $\langle\langle\ ,\ \rangle\rangle_0$ is positive definite on M_0^{+} (i.e., $\langle\langle n^{+}, n^{+}\rangle\rangle_0$ is positive for any $n^{+} \neq 0$ in M_0^{+}) and $\langle\langle\ ,\ \rangle\rangle_0$ is negative definite on M_0^{-} (i.e., $\langle\langle n^{-}, n^{-}\rangle\rangle_0$ is negative for any $n^{-} \neq 0$ in M_0^{-}). (See [AW, Theorem 6.2.51] or [HK, Section 10.2].) Since $\langle\langle\ ,\ \rangle\rangle_0$ is G-invariant, this is a decomposition as (real) representations of G. Thus to complete the proof we need only show that neither M_0^{+} nor M_0^{-} is all of M.

To do so, consider n_1. If $\langle\langle n_1, n_1\rangle\rangle_0 = 0$, we are done. Otherwise, by (7.1) and (7.2), $\langle\langle n_1, n_1\rangle\rangle_0$ and $\langle\langle n'_1, n'_1\rangle\rangle_0$ have opposite signs, and again we are done. $\qquad\square$

Corollary 7.10. *Let σ be an irreducible complex representation of G, $\sigma : G \to \mathrm{Aut}(M)$, with character χ. Then*

(1) $\mathrm{fs}(\sigma) = 0$ *if and only if χ is not real valued.*

(2) $\mathrm{fs}(\sigma) = 1$ *if and only if χ is real valued and M is defined over $\mathbf{R}$.*

(3) $\mathrm{fs}(\sigma) = -1$ *if and only if χ is real valued and M is not defined over $\mathbf{R}$.*

Proof. Immediate from Propositions 7.8 and 7.9. $\qquad\square$

Corollary 7.11. *Let σ be an irreducible complex representation of G, $\sigma : G \to \mathrm{Aut}(M)$, whose character χ is real valued. If the degree of M is odd, then σ is defined over $\mathbf{R}$.*

Proof. We present two separate proofs. Let d be the degree of M.

First proof: We know that $\sigma \oplus \bar{\sigma} = 2\sigma$ is defined over $\mathbf{R}$ (by Proposition 2.4.22) and that $d\sigma$ is defined over $\mathbf{R}$ (by Corollary 6.4). But $\gcd(2, d) = 1$ as d is odd, so, by Proposition 2.4.17, $1\sigma = \sigma$ is defined over $\mathbf{R}$.

Second proof: Suppose $\mathrm{fs}(\sigma) = -1$. Then M admits a nonsingular skew-symmetric bilinear form, so d is even. (Proof: Let A be the matrix of the form. Since A is nonsingular, $\det(A) \neq 0$, and since the form is skew symmetric, so is A, i.e., $A = -A^t$. Then $0 \neq \det(A) = \det(-A^t) = (-1)^d \det(A)$ so d is even.) The result follows by contraposition. $\qquad\square$

Example 7.12.

(1) Let ρ be the irreducible 2-dimensional complex representation of Q_8 constructed in Example 4.2. Direct computation shows that $\mathrm{fs}(\rho) = -1$, so ρ is not defined over $\mathbf{R}$, although χ_ρ is real valued.

Here is a direct proof that ρ is not defined over $\mathbf{R}$, though 2ρ is.

Suppose that ρ is defined over $\mathbf{R}$. Let $S = \rho(\mathbf{i})$ and $T = \rho(\mathbf{j})$. Since $S^4 = 1$, the minimum polynomial $m_S(X)$ of S must divide $X^4 - 1$. As S has order 4, and ρ has degree 2, we must have $m_S(X) = X^2 + 1$, and then also the characteristic polynomial $c_S(X)$ is $X^2 + 1$. Then S is similar to $C(X^2 + 1)$, the companion matrix of $X^2 + 1$, which is $\left[\begin{smallmatrix} 0 & -1 \\ 1 & 0 \end{smallmatrix}\right]$, and so we may assume that the matrix of S is $\left[\begin{smallmatrix} 0 & -1 \\ 1 & 0 \end{smallmatrix}\right]$. Let T have matrix $\left[\begin{smallmatrix} a & b \\ c & d \end{smallmatrix}\right]$. Now $\mathbf{ij} = -\mathbf{ji}$ in Q_8, so $ST = -TS$, and this equation yields $d = -a$, $c = b$. But T also has order 4, and direct computation shows that this is impossible if a and b are real.

On the other hand, if we replace each complex number $p + iq$ in the matrices of ρ by the two-by-two matrix $\left[\begin{smallmatrix} p & q \\ -q & p \end{smallmatrix}\right]$, we obtain a set of four-by-four real matrices, i.e., a real 4-dimensional representation of Q_8, and it is easy to check that this representation is equivalent to 2ρ.

(2) Recall that we gave the irreducible complex representations of the dihedral groups D_{2m} in Example 1.4 (7). From that we see immediately that all of the 1-dimensional irreducible representations of D_{2m} are defined over $\mathbf{R}$, and indeed over $\mathbf{Q}$. Let φ_k be an irreducible 2-dimensional complex representation of D_{2m}. Direct computation shows that $\mathrm{fs}(\varphi_k) = 1$, so φ_k is defined over $\mathbf{R}$. Note that, in the notation of Example 1.4 (7), the matrix of $\varphi_k(x)$ is not real valued. However, if we define the representation φ'_k of D_{2m} by

$$[\varphi'_k(x)] = \begin{bmatrix} \cos(2\pi ik/m) & \sin(2\pi ik/m) \\ -\sin(2\pi ik/m) & \cos(2\pi ik/m) \end{bmatrix} \quad \text{and} \quad [\varphi'_k(y)] = \begin{bmatrix} 0 & 1 \\ 1 & 0 \end{bmatrix},$$

then φ'_k is visibly defined over $\mathbf{R}$, and indeed over $\mathbf{Q}$ in case $m = 4$, and it is easy to check that φ_k and φ'_k are equivalent.

We close this section with the following application of the Frobenius-Schur indicator.

Proposition 7.13. *Let $\chi_1, \ldots, \chi_t$ be the characters of the irreducible complex representations of G, of degrees $d_1, \ldots, d_t$. Then*

(1) *For any $h \in G$, $\sum_{i=1}^{t} \mathrm{fs}(\chi_i)\chi_i(h) = $ the number of solutions of $g^2 = h$ in G.*

(2) $\sum_{i=1}^{t} \mathrm{fs}(\chi_i)d_i = 1 + $ *the number of elements of order 2 in G.*

Proof. Let $q(h)$ be the number of solutions of $g^2 = h$ in G, and note that $q(h) = q(h^{-1})$ and that $q(h) = q(k)$ if h and k are conjugate. Observe that

$$\mathrm{fs}(\chi_i) = \frac{1}{n} \sum_{g \in G} \chi_i(g^2)$$

$$= \frac{1}{n} \sum_{k \in G} \sum_{g^2 = k^{-1}} \chi_i(k^{-1})$$

$$= \frac{1}{n} \sum_{k \in G} q(k^{-1})\chi_i(k^{-1}).$$

Let h be in conjugacy class C_j. Then

$$\sum_{i=1}^{t} \mathrm{fs}(\chi_i)\chi_i(h) = \sum_{i=1}^{t} \frac{1}{n} \sum_{k \in G} q(k^{-1})\chi_i(k^{-1})\chi_i(h)$$

$$= \frac{1}{n} \sum_{k \in G} q(k^{-1}) \sum_{i=1}^{t} \chi_i(h)\chi_i(k^{-1})$$

$$= \frac{1}{n}|C_j|q(k^{-1})\frac{n}{|C_j|}$$

$$= q(k^{-1}) = q(k),$$

where the third equality follows from Corollary 5.18, proving (1).

Then (2) follows by setting $h = 1$. $\qquad\square$

Corollary 7.14. *If G has even order, at least one nontrivial irreducible complex representation of G is defined over $\mathbf{R}$.*

Proof. By Proposition 7.13 (2), $\mathrm{fs}(\chi_i) = 1$ for at least one nontrivial character of G. $\qquad\square$

3.8. Fields of Positive Characteristic

In section 3.6 we saw, among other things, how to obtain representations in characteristic p from those in characteristic 0. Our method was indirect: start with an irreducible representation in characteristic 0, obtain the corresponding central idempotent, reduce that idempotent mod p, and then obtain the corresponding irreducible representation in characteristic p. Here we shall see how to do this directly. The results of this section will also be of importance for us later, in our study of modular representations.

We begin by fixing a prime p.

Let $\mathbf{E}$ be an algebraic number field with ring of integers $\mathcal{O}$. Let $\mathcal{P}$ be a prime ideal of $\mathcal{O}$ lying above the ideal $p\mathbb{Z}$ of $\mathbb{Z}$. Then $\mathcal{P} \cap \mathbb{Z} = p\mathbb{Z}$ and $\mathcal{O}/\mathcal{P}$ is a finite field $\mathbf{F}$ of characteristic p. If $a \in \mathcal{O}$, we let $[a]$ be its image in $\mathcal{O}/\mathcal{P}$. We note that the map $a \mapsto [a]$ extends to the ring I of $\mathcal{P}$-integers of $\mathbf{E}$, i.e., to the elements of $\mathbf{E}$ of the form a/b, where $a \in \mathcal{O}$, $b \in \mathcal{O}$, and $b \notin \mathcal{P}$, by $a/b \mapsto [a/b] = [a][b]^{-1}$. We shall say that a $\mathcal{P}$-integer of $\mathbf{E}$ is divisible by $\mathcal{P}$ if it is of the form a/b, as above, with $a \in \mathcal{P}$, and write $a/b \equiv 0 \pmod{\mathcal{P}}$ in this case. Note that the ring I is isomorphic to the localization of the ring $\mathcal{O}$ at the prime ideal $\mathcal{P}$ ([AW, Example 2.3.6 (3)] or [Ma, Section 4]).

Construction 8.1 (Speiser). We present two variants of this construction.

Let σ be an irreducible complex representation of G of degree d, and recall that d necessarily divides n. Recall further that we may assume that σ is defined over $\overline{\mathbf{Q}}$. Choose a basis for the representation space, so that σ is given by matrices $\sigma(g) = Z(g) = (z_{ij}(g))$.

(1) By Corollary 2.4.25 (2), we may assume that all of the matrix entries of $\sigma(g)$ are algebraic integers, for every $g \in G$, and these finitely many algebraic integers generate a finite extension $\mathbf{E}$ of $\mathbf{Q}$.

(2) The matrix entries of $\sigma(g)$, for every $g \in G$, form a finite set, and so generate a finite extension $\mathbf{E}$ of $\mathbf{Q}$. Choose an ideal $\mathcal{P}$ of the ring of integers $\mathcal{O}$ of $\mathbf{E}$ as above, and form the ring I. As $\mathcal{O}$ is the ring of integers in the algebraic number field $\mathbf{E}$, $\mathcal{O}$ is a Dedekind domain ([Sa, Theorem 3.4.1]), and then I is a discrete valuation ring ([Ma, Section 11]). Then, by Corollary 2.4.25 (3), we may assume that σ is given by matrices $\sigma(g) = Z(g) = (z_{ij}(g))$ such that all of the entries of $\sigma(g)$ are $\mathcal{P}$-integers in $\mathbf{E}$, for every $g \in G$. (Note that [Sa, Proposition 5.1.4] proves that I is a PID, so using that, we may appeal to Proposition 2.4.24 directly.)

In either case, using the notation we have just introduced, we let $\overline{\sigma} = \mathrm{Red}_p(\sigma)$, the mod p reduction of σ, be the representation of G in characteristic p given by $\overline{\sigma}(g) = \overline{Z}(g) = ([z_{ij}(g)])$.

Remark. Note that in variant (1) above we have *not* claimed, and in general it is *not* true, that if $\mathbf{E} \subset \overline{\mathbf{Q}}$ is *any* field over which σ is defined, then all of the matrix entries of $\sigma(g)$, for every $g \in G$, can be chosen to be algebraic integers *in* $\mathbf{E}$.

Theorem 8.2 (Speiser). *If σ is an irreducible representation of degree d, and p is prime to n/d, then $\overline{\sigma} = \mathrm{Red}_p(\sigma)$ is irreducible.*

Proof. Suppose that $\overline{\sigma}$ is not irreducible. Then there is a nonsingular matrix $\overline{M} = (\overline{m}_{ij})$ such that $\overline{M}^{-1}\overline{Z}(g)\overline{M}$ is of the form

$$\overline{M}^{-1}\overline{Z}(g)\overline{M} = \begin{bmatrix} \overline{W}_1(g) & \overline{W}_2(g) \\ 0 & \overline{W}_4(g) \end{bmatrix}$$

for each $g \in G$. Let $M = (m_{ij})$ be any matrix with $[m_{ij}] = \overline{m}_{ij}$. Since $\overline{M}$ is nonsingular, so is M, and then $\det(M)$ is a $\mathcal{P}$-integer of $\mathbf{E}$, so

$$M^{-1}Z(g)M = \begin{bmatrix} W_1(g) & W_2(g) \\ W_3(g) & W_4(g) \end{bmatrix}$$

with $[W_i(g)] = \overline{W}_i(g)$. In particular, the entries of the submatrix $W_3(g)$ are $\mathcal{P}$-integers of $\mathbf{E}$. Let this whole d-by-d matrix have entries $(w_{ij}(g))$. Then by orthogonality of matrix coefficients (Proposition 5.6) we have

$$\sum_{g \in G} w_{d1}(g)w_{1d}(g^{-1}) = n/d.$$

But $w_{d1}(g) \equiv 0 \pmod{\mathcal{P}}$ for every $g \in G$, so $n/d \equiv 0 \pmod{\mathcal{P}}$, so, as n/d is an (ordinary) integer, $n/d \equiv 0 \pmod{p}$, a contradiction. $\quad\square$

Note that Speiser's construction involved a choice of matrix representation. We would like to know that $\mathrm{Red}_p(\sigma)$ is independent of that choice.

Corollary 8.3. *Let p be prime to n.*

(1) *For any irreducible complex representation σ of G, the isomorphism class of $\mathrm{Red}_p(\sigma)$ is well defined.*

(2) *If σ_1 and σ_2 are distinct irreducible complex representations of G, then $\mathrm{Red}_p(\sigma_1)$ and $\mathrm{Red}_p(\sigma_2)$ are distinct irreducible $\overline{\mathbf{F}}$-representations of G.*

(3) *As σ ranges over the distinct irreducible complex representations of G, $\mathrm{Red}_p(\sigma)$ ranges over the distinct irreducible $\overline{\mathbf{F}}$-representations of G.*

Proof. The trace of $\sigma(g)$ is well defined, and the trace of $\overline{\sigma}(g)$ is the mod p reduction of the trace of $\sigma(g)$, i.e., the character of $\overline{\sigma}(g)$ is the mod p reduction of the character of $\sigma(g)$.

The result then follows from the orthogonality of characters. $\quad\square$

Chapter 4

Induced Representations and Applications

4.1. Induced Representations

In this section we develop an important and powerful method of constructing representations—the technique of induction. First, however, we consider restriction.

In this section **F**, *G, and H are completely arbitrary. Any restrictions will be explicitly stated.*

Definition 1.1. Let G be a group, H a subgroup of G, and V an $\mathbf{F}(G)$-module. The **restriction of V to H**, $\mathrm{Res}_H^G(V)$, is V as an $\mathbf{F}(H)$-module.

In other words, we have the same underlying vector space, but we restrict the action to the subring $\mathbf{F}(H)$ of $\mathbf{F}(G)$. (We may sometimes simply write V for $\mathrm{Res}_H^G(V)$ when we wish to emphasize that the representation space is the same and it is clear from the context what group is operating.)

Let V be a representation of G and $H \subseteq G$ a subgroup. Clearly, if $\mathrm{Res}_H^G(V)$ is irreducible, then so is V, but the converse need not hold. For example, if $H = \{1\}$, $\mathrm{Res}_H^G(V) = \deg(V)\tau$. As a nontrivial example of restriction, let $G = D_{2m}$ and $H = \mathbb{Z}/m\mathbb{Z}$. Then, in the notation of Example 3.1.4 (6) and (7),

$$\mathrm{Res}_H^G(\varphi_i) = \theta_i \oplus \theta_{m-i}.$$

Example 1.2. Let H be a subgroup of G. Then

$$\operatorname{Res}_H^G(\mathbf{F}(G)) = [G : H]\mathbf{F}(H),$$

since $\mathbf{F}(G)$ is a free left $\mathbf{F}(H)$-module of rank $[G : H]$ (with a basis given by a set of right coset representatives).

Now we come to induction.

Definition 1.3. Let G be a group, H a subgroup of G, and W an $\mathbf{F}(H)$-module. Then the **induction of W to G**, denoted $\operatorname{Ind}_H^G(W)$, is given by

$$\operatorname{Ind}_H^G(W) = \mathbf{F}(G) \otimes_{\mathbf{F}(H)} W.$$

Note that this makes sense as we may regard $\mathbf{F}(G)$ as an $(\mathbf{F}(G), \mathbf{F}(H))$-bimodule, and then the result of induction is an $\mathbf{F}(G)$-module. If $V = \operatorname{Ind}_H^G(W)$, we say that V **is induced from** W and call V an **induced representation**.

Lemma 1.4.

(1) $\deg(\operatorname{Ind}_H^G(W)) = [G : H] \deg(W)$.

(2) W *is a direct summand of* $\operatorname{Res}_H^G \operatorname{Ind}_H^G(W)$.

(3) *If H has a set of left coset representatives $\{g_i\}$ with $H \subseteq C(g_i)$ for each i, then* $\operatorname{Res}_H^G \operatorname{Ind}_H^G(W) = [G : H]W$. *(Here $C(g_i)$ denotes the centralizer of g_i, i.e., $C(g_i) = \{g \in G \mid gg_i = g_i g\}$.)*

Proof.

(1) As a right $\mathbf{F}(H)$-module, or equivalently as an $(\mathbf{F}, \mathbf{F}(H))$-bimodule, $\mathbf{F}(G)$ is free of rank $[G : H]$ (with basis given by a set of left coset representatives).

(2) $\mathbf{F}(H)$ is a direct summand of $\mathbf{F}(G)$ as an $(\mathbf{F}(H), \mathbf{F}(H))$-bimodule, so $W = \mathbf{F}(H) \otimes_{\mathbf{F}(H)} W$ is a direct summand of $\mathbf{F}(G) \otimes_{\mathbf{F}(H)} W = \operatorname{Ind}_H^G(W)$ as a left $\mathbf{F}(H)$-module.

(3) In this case, $\mathbf{F}(G)$ is a free $(\mathbf{F}(H), \mathbf{F}(H))$-bimodule of rank $[G : H]$. $\square$

Remark. Observe that the hypothesis of Lemma 1.4 (3) implies that H is a normal subgroup of G. Observe also that it is satisfied if H is a subgroup of $Z(G)$, the center of G, or if G is the direct product $H \times K$ of H and another subgroup K.

Example 1.5. Let G be an arbitrary group and H a subgroup of G. Then for the regular representation $\mathbf{F}(H)$ of H, we have

$$\operatorname{Ind}_H^G(\mathbf{F}(H)) = \mathbf{F}(G),$$

the regular representation of G. This is immediate, for it is just the equality

$$\mathbf{F}(G) \otimes_{\mathbf{F}(H)} \mathbf{F}(H) = \mathbf{F}(G).$$

In particular, setting $H = \{1\}$, we have $\mathrm{Ind}_{\{1\}}^{G}(\tau) = \mathbf{F}(G)$.

Let us now give a "recognition principle" for induced representations.

Theorem 1.6. *Let $V = \mathrm{Ind}_{H}^{G}(W)$ and identify W with*

$$W_1 = \mathbf{F}(H) \otimes_{\mathbf{F}(H)} W \subseteq \mathbf{F}(G) \otimes_{\mathbf{F}(H)} W = V.$$

Let $\{g_i\}_{i \in I}$ be a complete set of left coset representatives of H in G, with $g_1 = 1$, and let $W_i = g_i(W_1)$. Then

(1) $V = \bigoplus_{i \in I} W_i$;

(2) *the action of G on V permutes the W_i, i.e., for every $g \in G$ and for every $i \in I$, $g(W_i) = W_j$ for some $j \in I$;*

(3) *this permutation is transitive, i.e., for every i, j, there is a $g \in G$ with $g(W_i) = W_j$; and*

(4) $H = \{g \in G : g(W_1) = W_1\}$.

Conversely, let V be an $\mathbf{F}(G)$-module and suppose there are subspaces $\{W_i\}_{i \in I}$ of V such that (1), (2), and (3) hold. Define H by (4), and set $W = W_1$. Then

$$V = \mathrm{Ind}_{H}^{G}(W).$$

Proof The direct statement of the proposition is clear from the isomorphism

$$\mathbf{F}(G) = \bigoplus_{i \in I} g_i \mathbf{F}(H)$$

of right $\mathbf{F}(H)$-modules. As for the converse, note that there is a one-to-one correspondence

$$\{W_i\} \longleftrightarrow \{\text{left cosets of } H\}$$

given by

$$W_i \longleftrightarrow \{g \in G : g(W_1) = W_i\}.$$

Pick coset representatives $\{g_i\}_{i \in I}$ with $g_1 = 1$. Define a function $\alpha : G \times W \to V$ by $\alpha(g, w) = g(w)$. This clearly extends to an $\mathbf{F}$-linear transformation $\alpha : \mathbf{F}(G) \times W \to V$, and since

$$\alpha(gh, w) = gh(w) = g(hw) = \alpha(g, hw),$$

it readily follows that α is $\mathbf{F}(H)$-middle linear and so defines

$$\widetilde{\alpha} : \mathbf{F}(G) \otimes_{\mathbf{F}(H)} W \to V.$$

Now we define $\beta : V \to \mathbf{F}(G) \otimes_{\mathbf{F}(H)} W$. Since $V = \bigoplus W_i$, it suffices to define $\beta : W_i \to \mathbf{F}(G) \otimes_{\mathbf{F}(H)} W$ for each $i \in I$. For $w_i \in W_i$ we let $\beta(w_i) = g_i \otimes g_i^{-1}(w_i)$. We shall check that $\widetilde{\alpha}$ and β are inverses of each other, establishing the claimed isomorphism.

First,

$$\widetilde{\alpha}\beta(w_i) = \widetilde{\alpha}(g_i \otimes g_i^{-1}(w_i)) = g_i(g_i^{-1}(w_i)) = w_i.$$

Second, each $g \in G$ can be written as $g = g_i h$ for a unique $i \in I$ and $h \in H$. Thus, for $w \in W$,

$$
\begin{aligned}
\beta\widetilde{\alpha}(g \otimes w) &= \beta(g(w)) \\
&= \beta(g_i h(w)) \\
&= \beta(g_i(hw)) \\
&= g_i \otimes g_i^{-1}(g_i(hw)) \\
&= g_i \otimes hw \\
&= g_i h \otimes w \\
&= g \otimes w,
\end{aligned}
$$

as required. $\qquad\qquad\square$

We make two observations. First, if V itself is irreducible, the condition of transitivity is automatic (for $\sum_{g \in G} g(W_1)$ is a subrepresentation of V, which must then be V itself). Second, our choice of W_1 was arbitrary; we could equally well have chosen some other W_i.

This second observation yields the following useful result.

Corollary 1.7. *Let H be a subgroup of G. Let $g_0 \in G$ be fixed and set $K = g_0^{-1} H g_0$. Let $\sigma_H : H \to \mathrm{Aut}(W)$ be a representation of H, and let $\sigma_K : K \to \mathrm{Aut}(W)$ be the representation of K defined by $\sigma_K(g) = \sigma_H(g_0 g g_0^{-1})$ for $g \in K$. Then*

$$\mathrm{Ind}_H^G(\sigma_H) = \mathrm{Ind}_K^G(\sigma_K).$$

Proof. Let $V = \mathrm{Ind}_H^G(W)$. Then we may identify W with W_1 in the statement of Theorem 1.6. Then $K = \{g \in G : g(W) = W\}$ acting on W by the above formula, so $V = \mathrm{Ind}_K^G(W)$ as well. $\qquad\square$

As a corollary of Theorem 1.6, we may identify two types of representations as induced representations.

Recall from Example 3.1.4 (8) that a representation $\sigma : G \to \mathrm{Aut}(V)$ is a permutation representation if V has a basis in which the matrix of $\sigma(g)$ is a permutation matrix, for every $g \in G$. (A permutation matrix has a single nonzero entry in every row and column, and that entry is equal to 1.) It is a monomial representation if V has a basis in which the matrix of $\sigma(g)$ has a single nonzero entry in every row and column, for every $g \in G$.

Corollary 1.8.

(1) *Let V be a transitive permutation representation on the set $P = \{p_i\}_{i \in I}$ and let $H = \{g \in G \mid g(p_1) = p_1\}$. Then $V = \mathrm{Ind}_H^G(\tau)$.*

(2) $\mathrm{Ind}_H^G(\tau)$ *is the natural permutation representation of G on left cosets of G/H (given by $g(g_0 H) = (g g_0)H$).*

(3) *Let V be a transitive monomial representation with respect to the basis $\mathcal{B} = \{b_i\}$, and let $H = \{g \in G \mid g(\mathbf{F}b_1) = \mathbf{F}b_1\}$. Let $\sigma = \mathbf{F}b_1$ as a representation of H. Then $V = \mathrm{Ind}_H^G(\sigma)$.*

Proof. $\square$

We will be investigating $\mathrm{Res}_H^G \mathrm{Ind}_H^G(W)$ in considerable detail below. However, we already know by Lemma 1.4 (2) that W is a subrepresentation of $\mathrm{Res}_H^G \mathrm{Ind}_H^G(W)$. This observation is enough to enable us to identify some induced representations.

For example, if $G = D_{2m}$ and φ_i is one of the 2-dimensional representations of G defined in Example 3.1.4 (7), then each φ_i, which is a monomial representation, is induced from $H = \mathbb{Z}/m\mathbb{Z}$, a subgroup of G of index two. If θ_i is the representation of that name of H in Example 3.1.4 (6), $i \leq m/2$, then $\mathrm{Ind}_H^G(\theta_i) = \varphi_i$, and if $i > m/2$, $\mathrm{Ind}_H^G(\theta_i) = \varphi_{m-i}$. This also illustrates Corollary 1.7, for if $g = y$ (in the notation of Example 3.1.4 (7)) and $\sigma_H = \theta_i$, then $K = H$, $\sigma_K = \theta_{m-i}$, and $\mathrm{Ind}_H^G(\theta_i) = \mathrm{Ind}_H^G(\theta_{m-i})$. (Of course, $\mathrm{Res}_H^G(\varphi_i) = \theta_i \oplus \theta_{m-i}$.)

Let us now concentrate on the case of a normal subgroup H of G and a representation $\sigma : H \to \mathrm{Aut}(W)$. If σ' is defined by $\sigma'(h) = \sigma(g^{-1}hg)$ for some fixed $g \in G$, we call σ' a **conjugate** of σ, or more precisely, the **conjugate of σ by g**.

Corollary 1.9. *Let H be a normal subgroup of G and $\sigma : H \to \mathrm{Aut}(W)$ a representation of H. Then if σ' is any conjugate of σ,*

$$\mathrm{Ind}_H^G(\sigma') = \mathrm{Ind}_H^G(\sigma).$$

Proof. This is a special case of Corollary 1.7. $\square$

Let $\{\sigma^j\}$ be a complete set of conjugates of $\sigma = \sigma^1$. Note that if we let

$$N(\sigma) = \{\, g \in G \mid \sigma' : H \to \mathrm{Aut}(W) \text{ defined by } \sigma'(h) = \sigma(g^{-1}hg)$$
$$\text{is a representation of } H \text{ isomorphic to } \sigma\},$$

then $N(\sigma)$ is a subgroup of G containing H, and $[G : N(\sigma)]$ is the number of conjugates of σ. (Note also that all the subgroups $N(\sigma^j)$ are mutually conjugate.) The subgroup $N(\sigma)$ is known as the **inertia group of** σ.

Corollary 1.10. *Let H be a normal subgroup of G and $\sigma : H \to \mathrm{Aut}(W)$ a representation of H. Let $\{\sigma^j\}$ be a complete set of conjugates of $\sigma = \sigma^1$. Then*

$$\mathrm{Res}_H^G \mathrm{Ind}_H^G(\sigma) = [N(\sigma) : H] \oplus_j \sigma^j.$$

Proof. Let $\{g_i\}$ be a set of coset representatives of H, with $g_1 = 1$, and let $\sigma_i : H \to \mathrm{Aut}(W)$ be defined by $\sigma_i(h) = \sigma(g_i^{-1} h g_i)$. Then

$$\mathrm{Res}_H^G \mathrm{Ind}_H^G(\sigma) = \bigoplus_i \sigma_i$$

by Theorem 1.6. But in the statement of the corollary we have just grouped the σ_i into isomorphism classes, there being $[N(\sigma) : H]$ of these in each class. $\qquad\square$

Theorem 1.11 (Clifford). *Let H be a normal subgroup of G. Let $\rho : G \to \mathrm{Aut}(V)$ be an irreducible representation of G, and let $\sigma : H \to \mathrm{Aut}(W)$ be any irreducible component of $\mathrm{Res}_H^G(\rho)$. Let $\{\sigma^j\}$ be a complete set of conjugates of $\sigma = \sigma^1$, and set $V_j = \sum_{g \in G} g(W)$, where the sum is taken over the left coset representatives of H such that the conjugate of σ by g is isomorphic to σ^j. Set $K = \{g \in G : g(V_1) = V_1\}$. Then V_1 is an irreducible representation of K and*

$$V = \mathrm{Ind}_K^G(V_1).$$

Furthermore, $K = N(\sigma)$.

Proof. First consider $V' = \sum_{g \in G} g(W)$. This is an $\mathbf{F}(G)$-submodule of V, but V is irreducible, so $V' = V$. Next observe that instead of summing over $g \in G$, we may instead sum over left coset representatives of H. Further note that the representation of H on $g(W)$ is the conjugate of σ by g^{-1}, so we may certainly group the terms together into isomorphism classes to get $V = \sum_j V_j$.

Now each V_j is a sum of subspaces, all of which are isomorphic to σ^j as an $\mathbf{F}(H)$-module, so by Lemma 2.2.5, each V_j is in fact isomorphic to a direct sum of these.

We claim that $V = \bigoplus_j V_j$. Consider $U = V_1 \cap \sum_{j>1} V_j$. Then U is an $\mathbf{F}(H)$-submodule of V_1, and so is isomorphic to $m_1 \sigma^1$ for some m_1. Also, U is an $\mathbf{F}(H)$-submodule of $\sum_{j>1} V_j$, and so is isomorphic to $\bigoplus_{j>1} m_j \sigma^j$ for some $\{m_j\}$. By Theorem 2.2.2, $m_1 = m_2 = \cdots = 0$, so $U = \{0\}$, as required.

If σ^j is the conjugate of σ by $g_j^{-1} \in G$, then $g_j(V_1) = V_j$, so G permutes the subspaces V_j transitively. Then, by Theorem 1.6, we obtain

$$V = \mathrm{Ind}_K^G(V_1).$$

Also, V_1 is irreducible, for if it contained a nonzero proper subrepresentation V_1', then $\mathrm{Ind}_K^G(V_1')$ would be a nonzero proper subrepresentation of V, contradicting the irreducibility of V.

Finally, $g \in K$ if and only if the conjugate by g of W is isomorphic to W. But this is exactly the condition for $g \in N(\sigma)$. $\qquad\square$

Corollary 1.12. *Let H be a normal subgroup of G. Let $\rho : G \to \mathrm{Aut}(V)$ be an irreducible representation of G, and let $\sigma = \sigma^1 : H \to \mathrm{Aut}(W)$ be any irreducible component of $\mathrm{Res}_H^G(\rho)$. Then*

(1) *$\mathrm{Res}_H^G(\rho)$ is a semisimple representation of H.*

(2) *$\mathrm{Res}_H^G(\rho) = m(\bigoplus_j \sigma^j)$ for some $m \in \{1, 2, 3, \dots\} \cup \{\infty\}$, where $\{\sigma^j\}$ is a complete set of conjugates of σ^1.*

(3) *Let $H^1 = \{g \in G \mid g(W) = W\}$. Then $m \leq [N(\sigma) : H^1]$. (Note that $H \subseteq H^1 \subseteq N(\sigma)$.)*

Proof. (1) In the notation of the proof of Theorem 1.11, we have that, as an $\mathbf{F}(H)$-module, $V = \sum_{g \in G} g(W)$, so V is a sum of simple $\mathbf{F}(H)$-modules and hence is semisimple by Corollary 2.2.6.

(2) Write $\mathrm{Res}_H^G(\rho)$ as a direct sum of irreducible representations of H. Let $V' \subseteq V$ be the subspace that is the sum of those subspaces of V on which H acts by some conjugate of σ^1. Then V' is an $\mathbf{F}(G)$-submodule of V, and V is irreducible, so $V' = V$. Hence we can conclude $\mathrm{Res}_H^G(\rho) = \bigoplus_j m_j \sigma^j$ for some $\{m_j\}$. Let V_j and g_j be as in the proof of Theorem 1.11, for each j. Then

$$m_j \sigma^j \cong V_j \cong g_j(V_1) \cong g_j(m_1 \sigma^1) \cong m_1 g_j(\sigma^1) \cong m_1 \sigma^j,$$

so $m_j = m_1$ for each j by Theorem 2.2.2. Set $m = m_1$.

(3) Since $g(W) = W$ for $g \in H^1$, $V_1 = \sum_j g(W)$ where the summation is over left coset representatives of H^1 in $N(\sigma)$. $\qquad\square$

Remark. It is not true in general that m is equal to $[N(\sigma) : H^1]$ or even that m divides $[N(\sigma) : H^1]$. Here is an example: Let p be an odd prime, let $G = \mathbb{Z}/p\mathbb{Z}$ and let $H = \{1\}$. Let $V = \mathcal{R}_0$ be the augmentation ideal of $\mathbf{Q}(G)$, a representation of G of degree $p - 1$. (Recall from Example 3.1.7 that $\mathcal{R}_0$ is irreducible.) Then $\mathrm{Res}_H^G = (p-1)\tau$, so $m = p-1$. Let W_1 be any

1-dimensional subspace of V. Then $\sigma = \tau$ and $N(\sigma) = G$, while $H^1 = \{1\}$, so $[N(\sigma) : H^1] = p$.

As a consequence of Clifford's theorem we have the following result (a representation is called **isotypic** if it is equivalent to some multiple of a single irreducible representation):

Corollary 1.13. *Let H be a normal subgroup of G, and let $\rho : G \to \mathrm{Aut}(V)$ be an irreducible representation of G. If $\mathrm{Res}_H^G(V)$ is not isotypic, then there is a proper subgroup K of G containing H and an irreducible representation U of K such that $V = \mathrm{Ind}_K^G(U)$.*

Proof. Suppose $\mathrm{Res}_H^G(V)$ is not isotypic. Let W be an irreducible $\mathbf{F}(H)$-submodule of $\mathrm{Res}_H^G(V)$, and set $U = \sum_g g(W)$ where the sum is over all $g \in G$ with $g(W)$ isomorphic to W as an $\mathbf{F}(H)$-module, and let $K = \{g \in G \mid g(U) = U\}$. Then, as above, U is an irreducible $\mathbf{F}(K)$-module. Note $H \subseteq K$ and $K \subset G$ as $\mathrm{Res}_H^G(V)$ is not isotypic. Then $\mathrm{Ind}_K^G(U)$ is an $\mathbf{F}(G)$-submodule of V, but as V is irreducible, it is equal to V. $\qquad\square$

As an example of this corollary, let $m = m_1 m_2$ with $m_1 > 2$, let $G = D_{2m}$, $V = \varphi_1$, and $H = \mathbb{Z}/m_1\mathbb{Z}$. Then the corollary applies with $K = \mathbb{Z}/m\mathbb{Z}$ and $V_1 = \theta_1$. (Note that if $m_2 = 1$, then $K = H$, while if $m_2 > 1$, $K \supset H$.)

We may use Corollary 1.13 to sharpen Theorem 3.6.3.

Theorem 1.14. *Let G be a finite group and A an abelian normal subgroup of G. If V is an irreducible complex representation of G, then $d = \deg(V)$ divides $[G : A]$.*

Proof: We prove this by induction on $n = |G|$. Let V be defined by $\sigma : G \to \mathrm{Aut}(V)$. Let $W = \mathrm{Res}_A^G(V)$.

If W is not isotypic, then, by Corollary 1.13, $V = \mathrm{Ind}_K^G(U)$ for some proper subgroup K of G containing A and some irreducible representation U of K. Then, by induction, $\deg(U)$ divides $[K : A]$, so in this case $\deg(V) = [G : K]\deg(U)$ divides $[G : K][K : A] = [G : A]$ and we are done.

If W is isotypic, then $W = dW_1$ for some 1-dimensional representation W_1 of A, or, in other words, $\sigma \mid A \to \mathrm{Aut}(W) = \mathrm{Aut}(V)$ is given by $\sigma(g) =$ multiplication by some complex number $\lambda(g)$, for each $g \in A$. (Also, for any $h \in G$, $\lambda(hgh^{-1}) = \lambda(g)$, $g \in A$.)

First suppose that A is a subgroup of the center of G. For each m consider $\sigma^{\otimes m}$ on $V \otimes \cdots \otimes V$. It is easy to check that this is an irreducible representation of $G^m = G \times \cdots \times G$. Let

$$H = \{(g_1, \ldots, g_m) \in A^m \mid \sigma^{\otimes m}(g_1, \ldots, g_m) = \lambda(g_1)\cdots\lambda(g_m) = 1\}.$$

H is a subgroup of A^m and hence a normal subgroup of G^m, and it acts trivially on $V^{\otimes m}$, so we obtain an irreducible representation of G^m/H, which is irreducible as $\sigma^{\otimes m}$ is. Hence, by Theorem 3.6.3, the degree d^m of this representation divides the order of G^m/H.

Note, however, that H has a subgroup

$$\{(g_1, \ldots, g_{m-1}, g_1^{-1} \cdots g_{m-1}^{-1}) \mid g_i \in A\}$$

isomorphic to A^{m-1}, so if $a = |A|$, then $|H|$ is divisible by a^{m-1}, and so d^m divides $n^m/a^{m-1} = a(n/a)^m$ for every m, which implies that d divides $n/a = [G : A]$, as claimed.

Now for the general case.

Suppose that σ is not faithful, and let $K = \mathrm{Ker}(\sigma)$. Then $\sigma : G \to \mathrm{Aut}(V)$ factors through G/K, and the image of A in G/K is $A/A \cap K$. By induction, d divides $[G/K : A/A \cap K]$. But $A/A \cap K$ is isomorphic to AK/K, so $[G/K : A/A \cap K] = [G : K]/[AK : K] = [G : AK]$ which divides $[G : A]$.

Thus we are reduced to the case σ faithful. As σ is faithful, so is its restriction to A. But $\sigma(a)$ is multiplication by $\lambda(a)$ for each $a \in A$, so $\sigma(gag^{-1})$ is multiplication by $\lambda(gag^{-1}) = \lambda(a)$. Hence we must have $gag^{-1} = a$ for all $g \in G$, $\alpha \in A$, so A is in the center of G, and we are reduced to the first case. $\qquad\square$

Remark. The case A contained in the center of G is due to Schur, though this proof is due to Tate. The general case is due to Ito.

We now return to the general study of induction.

Lemma 1.15. *Let $K \subseteq H \subseteq G$ be subgroups.*

(1) (Transitivity of restriction) *For any representation V of G,*

$$\mathrm{Res}_K^H \mathrm{Res}_H^G(V) = \mathrm{Res}_K^G(V).$$

(2) (Transitivity of induction) *For any representation W of K,*

$$\mathrm{Ind}_H^G \mathrm{Ind}_K^H(W) = \mathrm{Ind}_K^G(W).$$

Proof. (1) is trivial. As for (2),

$$
\begin{aligned}
\mathrm{Ind}_H^G \mathrm{Ind}_K^H(W) &= \mathbf{F}(G) \otimes_{\mathbf{F}(H)} (\mathbf{F}(H) \otimes_{\mathbf{F}(K)} W) \\
&= (\mathbf{F}(G) \otimes_{\mathbf{F}(H)} \mathbf{F}(H)) \otimes_{\mathbf{F}(K)} W \\
&= \mathbf{F}(G) \otimes_{\mathbf{F}(K)} W \\
&= \mathrm{Ind}_K^G(W),
\end{aligned}
$$

where the second equality is just the associativity of the tensor product
([AW, theorem 7.2.17]). $\qquad\square$

The next formula turns out to be tremendously useful, and we will see
many examples of its use. Recall that we defined the intertwining number

$$i(V, W) \;=\; \langle V, W \rangle$$

of two representations in Definition 3.5.19.

Theorem 1.16 (Frobenius reciprocity). *Let $\mathbf{F}$ be an arbitrary field, G
an arbitrary group, and H a subgroup of G. Let W be an $\mathbf{F}$-representation
of H and V an $\mathbf{F}$-representation of G. Then*

$$\langle \operatorname{Ind}_H^G(W),\ V \rangle \;=\; \langle W,\ \operatorname{Res}_H^G(V) \rangle.$$

Proof. By definition,

$$\langle \operatorname{Ind}_H^G(W),\ V \rangle \;=\; \dim_{\mathbf{F}} \operatorname{Hom}_{\mathbf{F}(G)}(\operatorname{Ind}_H^G(W),\ V).$$

But

$$
\begin{aligned}
\operatorname{Hom}_{\mathbf{F}(G)}(\operatorname{Ind}_H^G(W), V) &\cong \operatorname{Hom}_{\mathbf{F}(G)}(\mathbf{F}(G) \otimes_{\mathbf{F}(H)} W, V) \\
&\cong \operatorname{Hom}_{\mathbf{F}(H)}(W, \operatorname{Hom}_{\mathbf{F}(G)}(\mathbf{F}(G), V)), \\
&\cong \operatorname{Hom}_{\mathbf{F}(H)}(W, V)
\end{aligned}
$$

where the second isomorphism is the adjoint associativity of Hom and tensor
product (Theorem A.1.4). Again, by definition,

$$\dim_{\mathbf{F}} \operatorname{Hom}_{\mathbf{F}(H)}(W, V) \;=\; \langle W, \operatorname{Res}_H^G(V) \rangle.$$

$$\square$$

One direct consequence of Frobenius reciprocity occurs so often that it
is worth stating explicitly.

Corollary 1.17. *Let G be a finite group and let H be a subgroup of G.
Let $\mathbf{F}$ be a field with $\operatorname{char}(\mathbf{F}) = 0$ or prime to the order of G. Let W
be an absolutely irreducible $\mathbf{F}$-representation of H and V be an absolutely
irreducible $\mathbf{F}$-representation of G. Then the multiplicity of V in $\operatorname{Ind}_H^G(W)$
is equal to the multiplicity of W in $\operatorname{Res}_H^G(V)$.*

Proof. Immediate from Theorem 1.16 and Lemma 3.5.21. $\qquad\square$

As an example of the use of Frobenius reciprocity let us use it to provide an alternate proof of part (2) of Frobenius's fundamental theorem,

Theorem 3.3.3. Let $\mathbf{F}$ be an excellent field for G and M an irreducible $\mathbf{F}$-representation of G of degree d. Let m be the multiplicity of M in $\mathbf{F}(G)$. We wish to show $m = d$.

Let $H = \{1\}$ and recall from Example 1.5 that $\mathbf{F}(G) = \mathrm{Ind}_H^G(\tau)$. Then

$$\begin{aligned}
m &= \langle M,\ \mathbf{F}(G)\rangle \\
&= \langle \mathbf{F}(G),\ M\rangle \\
&= \langle \mathrm{Ind}_H^G(\tau),\ M\rangle \\
&= \langle \tau,\ \mathrm{Res}_H^G(M)\rangle \\
&= \langle \tau,\ d\tau\rangle \\
&= d
\end{aligned}$$

as claimed.

Example 1.18. Consider A_4 from Example 3.4.3, and let $\mathbf{F} = \mathbf{C}$. We have a split extension

$$1 \longrightarrow V \longrightarrow A_4 \xrightarrow{\ \pi\ } S \longrightarrow 1$$

with $V \cong \mathbb{Z}/2\mathbb{Z} \oplus \mathbb{Z}/2\mathbb{Z}$. V has four irreducible representations, all of degree 1: τ, and three others, which we shall simply denote by λ_1, λ_2, and λ_3. A_4 has four irreducible representations: $\pi^*(\theta_i)$ for $i = 0, 1, 2$, of degree 1, and α, of degree 3. Then

$$\begin{aligned}
\langle \mathrm{Ind}_V^{A_4}(\tau),\ \pi^*(\theta_i)\rangle &= \langle \tau,\ \mathrm{Res}_V^{A_4}(\pi^*(\theta_i))\rangle \\
&= \langle \tau,\ \tau\rangle \\
&= 1,
\end{aligned}$$

so

$$\mathrm{Ind}_V^{A_4}(\tau) = \overset{2}{\underset{i=0}{\bigoplus}} \pi^*(\theta_i)$$

(since

$$\deg(\mathrm{Ind}_V^{A_4}(\tau)) = [A_4 : V]\deg(\tau) = 3 \cdot 1 = 3$$

and the right-hand side is a representation of degree 3). Also, for $i = 0, 1, 2$, and $j = 1, 2, 3$,

$$0 = \langle \lambda_j, \tau\rangle = \langle \lambda_j,\ \mathrm{Res}_H^G(\pi^*(\theta_i))\rangle = \langle \mathrm{Ind}_H^G(\lambda_j), \pi^*(\theta_i)\rangle$$

so we must have $\mathrm{Ind}_H^G(\lambda_j) = \alpha$ (as both are of degree 3).

Continuing with this example, since we have a split extension, we have a subgroup S of A_4 isomorphic to $\mathbb{Z}/3\mathbb{Z}$, and we identify S with $\mathbb{Z}/3\mathbb{Z}$ via this isomorphism. (We have given S and this isomorphism explicitly in Example

3.4.3). Now S has three irreducible representations $\theta_0 = \tau$, θ_1, and θ_2. Because we have a splitting, $\operatorname{Res}_S^{A_4}(\pi^*(\theta_i)) = \theta_i$, or, more generally,

$$
\begin{aligned}
\delta_{ij} &= \langle \operatorname{Res}_S^{A_4}(\pi^*(\theta_j)), \ \theta_i \rangle \\
&= \langle \pi^*(\theta_j), \ \operatorname{Ind}_S^{A_4}(\theta_i) \rangle;
\end{aligned}
$$

since $\operatorname{Ind}_S^{A_4}(\theta_i)$ has degree 4, this gives

$$
\operatorname{Ind}_S^{A_4}(\theta_i) = \pi^*(\theta_i) \oplus \alpha \quad \text{for } i = 0, \ 1, \ 2.
$$

Here is another useful consequence of Frobenius reciprocity.

Proposition 1.19. *Let G be a finite group and let $\mathbf{F}$ be a field with $\operatorname{char}(\mathbf{F}) = 0$ or prime to the order of G. Let $d_{\max}(G)$ denote the maximal degree of an absolutely irreducible $\mathbf{F}$-representation of G. Then for every subgroup H, $d_{\max}(G) \leq [G : H] d_{\max}(H)$. In particular, $d_{\max}(G) \leq [G : A]$ for every abelian subgroup A of G.*

Proof. Let V be an irreducible representation of G. Then $\operatorname{Res}_H^G(V)$ is a representation of H and so contains an irreducible representation W. Then

$$
1 \ \leq \ \langle \operatorname{Res}_H^G(V), \ W \rangle \ = \ \langle V, \ \operatorname{Ind}_H^G(W) \rangle
$$

(the equality being Frobenius reciprocity), so V is a subrepresentation of $\operatorname{Ind}_H^G(W)$. But $\deg(\operatorname{Ind}_H^G(W)) = [G : H] \deg(W)$, so we see that $\deg(V) \leq [G : H] \deg(W)$. $\qquad\qquad\square$

For example, D_{2m} has an abelian subgroup of index 2 and its irreducible complex representations all have dimension at most 2. The same is true for Q_8. Also, A_4 has an abelian subgroup of index 3 and its irreducible complex representations all have dimension at most 3.

Now we determine the character of an induced representation. In order to state this most simply, we adopt in this theorem and henceforth the following nonstandard notation. If H is a subgroup of G and W an $\mathbf{F}$-representation of H with character χ_W, we let $\widetilde{\chi}_W$ be the function on G defined by

$$
\widetilde{\chi}_W(g) = \begin{cases} \chi_W(g) & \text{if } g \in H, \\ 0 & \text{if } g \notin H. \end{cases}
$$

Theorem 1.20. *Let* $\mathbf{F}$ *be an arbitrary field, let* G *be an arbitrary group, and let* H *be a subgroup of* G *of finite index* k. *Let* W *be an* $\mathbf{F}$-*representation of* H *of finite degree. Set* $V = \operatorname{Ind}_H^G(W)$. *If* $\{g_i\}_{i=1}^k$ *is a complete set of left coset representatives of* H, *then for any* $g \in G$,

$$\chi_V(g) = \sum_{i=1}^k \widetilde{\chi}_W(g_i^{-1}gg_i).$$

Proof. We know that we may write

$$V = \bigoplus_{i=1}^k g_i(W) = \bigoplus_{i=1}^k W_i,$$

and that every element of G acts by permuting $\{W_i\}$. To be precise, if we let $H_i = g_i H g_i^{-1}$, then

$$H_i = \{g \in G : g(W_i) = W_i\},$$

so if $g \notin H_i$, then $g(W_i) = W_j$ for some $j \neq i$. On the other hand, if the representation of H_i on W_i is given by $\sigma_i : H_i \to \operatorname{Aut}(W_i)$, with $\sigma_1 = \sigma$, then for $g \in H_i$, $\sigma_i(g) = \sigma(g')$, where $g' = g_i^{-1}gg_i$. Now $f_i : W_1 \to W_i$ by $f_i(w) = g_i(w)$ is an isomorphism with

$$(f_i\sigma(g')f_i^{-1})(w_i) = \sigma_i(g)(w_i)$$

for all $g \in H_i$ and $w_i \in W_i$, so

$$\operatorname{Tr}(\sigma_i(g)) = \operatorname{Tr}(\sigma(g')).$$

Now $\chi_V(g)$ is the trace of a matrix representing the operation of g. Choose a basis for V that is a union of bases for the W_i. Then if $g \in G$ with $g \notin H_i$, the action of g on W_i contributes nothing to $\chi_V(g)$; while if $g \in H_i$, it contributes $\operatorname{Tr}(\sigma(g')) = \chi_W(g') = \chi_W(g_i^{-1}gg_i)$ to $\chi_V(g)$, yielding the theorem. $\qquad\square$

Corollary 1.21. *In the situation of Theorem 1.20, if* H *is finite as well and* $\operatorname{char}(\mathbf{F}) = 0$ *or is prime to the order of* H,

$$\chi_V(g) = \frac{1}{|H|} \sum_{g' \in G} \widetilde{\chi}_W(g'^{-1}gg').$$

Proof. $\qquad\square$

Corollary 1.22. *Let* $\mathbf{F}$, G, *and* H *be as in Theorem 1.20. Then:*

(1) $\chi_V(g) = 0$ *if* g *is not conjugate to an element of* H.

In particular, if H *is a normal subgroup of* G, $\chi_V(g) = 0$ *for* $g \notin H$.

(2) *If G is abelian or, more generally, if H is a subgroup of the center $Z(G)$ of G, then*

$$\chi_V(g) = \begin{cases} k\chi_W(g) & \text{if } g \in H, \\ 0 & \text{if } g \notin H. \end{cases}$$

(3) *If $W = \tau$ and C_g is finite, then*

$$|C_g|\chi_V(g) = [G:H]|C_g \cap H|,$$

where C_g denotes the conjugacy class of G containing g.

(4) *If H is cyclic and $W = \tau$, then*

$$\chi_V(g) = \begin{cases} [N(H_0):H] & \text{if } g \text{ is conjugate to } h_0 \in H, \\ 0 & \text{if } g \text{ is not conjugate to an element of } H, \end{cases}$$

where H_0 denotes the subgroup of H generated by h_0.

In particular, $\chi_V(g) = [N(H):H]$ if g is conjugate to a generator of H, and $\chi_V(g)$ is a multiple of $[N(H):H]$ for every $g \in G$. (Here $N(H) = \{g \in G \mid gHg^{-1} = H\}$ is the normalizer of the subgroup H.)

Proof. (1) and (2) are immediate.

As for (3), note that χ_V is constant on conjugacy classes, and also that, for any fixed g_i, as g' runs over C_g, so does $g_i^{-1}g'g_i$. Then, if $k = [G:H]$,

$$\begin{aligned}
|C_g|\chi_V(g) &= \sum_{g' \in C_g} \sum_{i=1}^{k} \tilde{\chi}_W(g_i^{-1}g'g_i) \\
&= \sum_{i=1}^{k} \sum_{g' \in C_g} \tilde{\chi}_W(g_i^{-1}g'g_i) \\
&= \sum_{i=1}^{k} \sum_{g' \in C_g} \tilde{\chi}_W(g') = k|C_g \cap H|
\end{aligned}$$

as $\tilde{\chi}_W(g') = 1$ if $g' \in H$ and 0 if not.

As for (4), note that a subgroup H_0 of a cyclic group H is uniquely determined by its index in H. Then, for any $g \in N(H)$, $gH_0g^{-1} \subseteq H$ and hence $gH_0g^{-1} = H_0$, so $g \in N(H_0)$. Thus $N(H) \subseteq N(H_0)$ and $k_0 = [N(H_0):H] = [N(H_0):N(H)][N(H):H]$ is a multiple of $[N(H):H]$.

Now, since characters are class functions, it suffices to prove (4) for $g = h_0 \in H$. Theorem 1.20 is valid for any set of coset representatives $\{g_i\}$, so we choose $\{g_i\}$ with $g_i \in N(H_0)$ for $i = 1, \dots, k_0$ and $g_i \notin N(H_0)$ for

$i > k_0$. Then

$$\chi_V(h_0) = \sum_{i=1}^{k} \widetilde{\chi}_W(g_i^{-1}g'g_i)$$

$$= |\{g_i \mid g_i^{-1}h_0g_i \in H\}| = |\{g_i \mid g_i^{-1}H_0g_i \subseteq H\}|$$

$$= |\{g_i \mid g_i^{-1}H_0g_i = H_0\}| = k_0$$

as claimed. $\square$

Remark. In the equality in Corollary 1.22 (3), it need not be the case that either of the two terms on one side of the equation divides either of the two terms on the other side (see Example 3.6 (1)).

Our definition of intertwining numbers of characters in Lemma 3.5.21 (5) clearly extends to class functions. The formula for χ_V in Theorem 1.20 also clearly extends to class functions, i.e., if f is a class function on H, then we define

$$\mathrm{Ind}_H^G(f)(g) = \sum_{i=1}^{k} \widetilde{f}(g_i^{-1}gg_i),$$

where $\widetilde{f}(g) = f(g)$ if $g \in H$ and $\widetilde{f}(g) = 0$ if $g \notin H$. Then, if $\mathbf{F}$ is a field of characteristic 0, and θ is a class function on H, Frobenius reciprocity gives

$$\langle \mathrm{Ind}_H^G(f), \theta \rangle = \langle f, \mathrm{Res}_H^G(\theta) \rangle$$

where $\mathrm{Res}_H^G(\theta)$ denotes the restriction of the class function θ to H.

Lemma 1.23. *Let f be a class function on H and θ be a class function on G. Then, as functions on G,*

$$\mathrm{Ind}_H^G(f)\theta = \mathrm{Ind}_H^G(f \, \mathrm{Res}_H^G(\theta)).$$

Proof.

$$(\mathrm{Ind}_H^G(f)\theta)(g) = \sum_{i=1}^{k} \widetilde{f}(g_i^{-1}gg_i)\theta(g)$$

$$= \sum_{i=1}^{k} \widetilde{f}(g_i^{-1}gg_i)\theta(g_i^{-1}gg_i)$$

$$= \sum_{i=1}^{k} \widetilde{f}(g_i^{-1}gg_i)\widetilde{\theta}(g_i^{-1}gg_i)$$

$$= \sum_{i=1}^{k} \widetilde{f\theta}(g_i^{-1}gg_i)$$

$$= \mathrm{Ind}_H^G(f \, \mathrm{Res}_H^G(\theta))(g).$$

$\square$

Theorem 1.24. *Let G be a finite group, let H be a subgroup of G, and let R be the ring of complex class functions on G,*

$$R = \{\theta : G \to \mathbf{C} \mid \theta \text{ is constant on conjugacy classes}\}.$$

Let $I \subseteq R$ be defined by

$$I = \{\operatorname{Ind}_H^G(f) \mid f \text{ is a complex class function on } H\}.$$

Then I is an ideal in R and coincides with the ideal

$$J = \{\theta \in R \mid \theta(g) = 0 \text{ if } g \text{ is not conjugate to an element of } H\}.$$

Proof. I is an ideal of R by Lemma 1.23, and $I \subseteq J$ by Corollary 1.22 (1). We need to show $J \subseteq I$. Let θ_C denote the characteristic function of the conjugacy class $C \subset G$. It suffices to show that $\theta_C \in I$ for each conjugacy class C with $C \cap H \neq \emptyset$.

Let $\emptyset \neq C \cap H = D_1 \cup \cdots \cup D_s$ be a decomposition of $C \cap H$ into H-conjugacy classes. Let D be any D_j and let f_D be the characteristic function of the conjugacy class D.

Let

$$\varphi_C = \frac{|C|}{|D|} \operatorname{Ind}_H^G(f_D).$$

Then $\varphi_C \in I$, and we claim $\theta_C = \varphi_C$. To show this, it suffices to show that

$$\langle \theta_C, \theta_{C'} \rangle = \langle \varphi_C, \theta_{C'} \rangle$$

for every conjugacy class C' of G.

Now

$$\langle \theta_C, \theta_{C'} \rangle = \frac{1}{n} \sum_{g \in G} \theta_C(g) \theta_{C'}(g) = \begin{cases} 0 & \text{if } C' \neq C, \\ |C|/n & \text{if } C' = C. \end{cases}$$

On the other hand

$$
\begin{aligned}
\langle \varphi_C, \theta_{C'} \rangle &= \langle \frac{|C|}{|D|} \operatorname{Ind}_H^G(f_D), \theta_{C'} \rangle \\
&= \langle \frac{|C|}{|D|} f_D, \operatorname{Res}_H^G(\theta_{C'}) \rangle \\
&= \frac{1}{n} \sum_{g \in G} \frac{|C|}{|D|} f_D(g) \operatorname{Res}_H^G(\theta_{C'})(g) \\
&= \begin{cases} 0 & \text{if } C' \neq C, \\ |C|/n & \text{if } C' = C, \end{cases}
\end{aligned}
$$

completing the proof. $\square$

Here is an alternate formula, due to Frobenius, for the character of an induced representation.

Theorem 1.25 (Frobenius). *Let G be a finite group and let H be a subgroup of G. Let W be an $\mathbf{F}$-representation of H, and set $V = \operatorname{Ind}_H^G(W)$. Let $c \in G$ be in conjugacy class C, and let $C \cap H = D_1 \cup \cdots \cup D_s$, a decomposition into H-conjugacy classes. Then*

$$|C|\chi_V(c) = [G : H] \sum_{j=1}^{s} |D_j|\chi_W(d_j) \quad \text{where } d_j \in D_j.$$

(If $C \cap H = \emptyset$, $s = 0$, the sum is empty, and its value is 0.)

Proof. By Theorem 1.20, if $\{g_i\}$ is a set of left coset representatives of H,

$$|C|\chi_V(c) = \sum_{c' \in C} \sum_{g_i} \widetilde{\chi}_W(g_i^{-1}c'g_i)$$

$$= \sum_{j=1}^{s} \delta_j \chi_W(d_j),$$

where

$$\begin{aligned}
\delta_j &= |\{(g_i, c') \mid g_i^{-1}c'g_i \in D_j\}| = |\{(g_i, c', d_j') \mid g_i^{-1}c'g_i = d_j', \ d_j' \in D_j\}| \\
&= |\{(g_i, d_j') \mid g_i^{-1}c'g_i = d_j'\}| = [G : H]|D_j|
\end{aligned}$$

as for any d_j' and any g_i, $g_i^{-1}c'g_i = d_j'$ has the unique solution $c' = g_i d_j' g_i^{-1}$ (and $c' \in C$). $\qquad\square$

Example 1.26. In Example 3.5.26, we found the character table of A_5. Here let us adopt an alternate approach, finding the irreducible characters via induced representations. We still know, of course, that they must have degrees 1, 3, 3, 4, 5 and we still denote them as in Example 3.5.26.

Of course, $\alpha_1 = \tau$. The construction of α_4 was so straightforward that an alternative is hardly necessary, but we shall give one anyway (as we shall need most of the work in any case). Let $G = A_5$ and let $H = A_4$ included in the obvious way (as permutations of $\{1, 2, 3, 4\} \subseteq \{1, 2, 3, 4, 5\}$). Then a system of left coset representatives for H is

$$\{1, (1\ 2)(4\ 5), (1\ 2)(3\ 5), (1\ 3)(2\ 5), (2\ 3)(1\ 5)\} = \{g_1, \ldots, g_5\}.$$

We choose as representatives for the conjugacy classes of G

$$\{1, (1\ 4)(2\ 3), (1\ 2\ 3), (1\ 2\ 3\ 4\ 5), (1\ 3\ 5\ 2\ 4)\} = \{c_1, \ldots, c_5\}.$$

Of course, $g_i^{-1}(c_1)g_i = c_1$ for every i. Otherwise, one can check that $g_i^{-1}(c_j)g_i \notin H$ except in the following cases:

$$g_1^{-1}(c_2)g_1 = c_2, \quad g_1^{-1}(c_3)g_1 = c_3, \quad g_2^{-1}((1\ 2\ 3))g_2 = (1\ 3\ 2).$$

(Note that $(1\ 2\ 3) = c_3$, and that while $(1\ 3\ 2)$ is conjugate to it in A_5, it is *not* conjugate to it in A_4, so we have written the permutation explicitly.)

Let $W = \tau$ and consider $\mathrm{Ind}_H^G(W) = V$. Then by Theorem 1.20, it is easy to compute χ_V:

$$\chi_V(c_1) = 5, \quad \chi_V(c_2) = 1, \quad \chi_V(c_3) = 2, \quad \chi_V(c_4) = \chi_V(c_5) = 0.$$

Then

$$\langle \chi_V, \chi_V \rangle \;=\; \frac{1}{60}(5^2 + 15 \cdot 1^2 + 20 \cdot 2^2) = 2,$$

and τ appears in V with multiplicity 1 by Frobenius reciprocity (or by calculating $\langle \chi_V, \chi_1 \rangle = 1$), so V contains one other irreducible representation, $V = \tau \oplus \alpha_4$, and $\chi_4 = \chi_V - \chi_1$, giving the character of α_4.

Now, following Example 3.4.3, let $W = \pi^*(\theta_1)$ (or $\pi^*(\theta_2)$) and consider $\mathrm{Ind}_H^G(W) = V$. Again, it is easy to compute χ_V: $\chi_V(c_1) = 5$, $\chi_V(c_2) = 1$, $\chi_V(c_3) = \exp(2\pi i/3) + \exp(4\pi i/3) = -1$, $\chi_V(c_4) = \chi_V(c_5) = 0$.

Now τ does not appear in V by Frobenius reciprocity, so that implies here (by considering degrees) that V is irreducible (or alternatively one may calculate that $\langle \chi_V, \chi_V \rangle = 1$), so $V = \alpha_5$ and its character is given above.

Now we are left with determining the characters of the two irreducible representations of degree 3. To find these, let $H = \mathbb{Z}/5\mathbb{Z}$ be the subgroup generated by the 5-cycle $(1\ 2\ 3\ 4\ 5)$. Then H has a system of left coset representatives

$$\{g_1, \ldots, g_{12}\} = \{1, \ (1\ 4)(2\ 3), \ (2\ 4\ 3), \ (1\ 4\ 2), \ (2\ 3\ 4), \ (1\ 4\ 3),$$
$$(1\ 2)(3\ 4), \ (1\ 3)(2\ 4), \ (1\ 2\ 3), \ (1\ 3\ 4), \ (1\ 2\ 4), \ (1\ 3\ 2)\}.$$

Again, $g_i^{-1}(c_1)g_i = c_1$ for every i. Otherwise, one can check that $g_i^{-1}(c_j)g_i \notin H$ except in the following cases:

$$g_1^{-1}(c_4)g_1 = c_4, \quad g_1^{-1}(c_5)g_1 = c_5,$$
$$g_2^{-1}((1\ 2\ 3\ 4\ 5))g_2 = (1\ 5\ 4\ 3\ 2) = (1\ 2\ 3\ 4\ 5)^{-1},$$
$$g_2^{-1}((1\ 3\ 5\ 2\ 4))g_2 = (1\ 4\ 2\ 5\ 3) = (1\ 3\ 5\ 2\ 4)^{-1}.$$

Now let $W = \theta_j$ and let $V = \mathrm{Ind}_H^G(W)$. Again by Theorem 1.20, we compute χ_V: $\chi_V(c_1) = 12$, $\chi_V(c_2) = \chi_V(c_3) = 0$, and

$$\chi_V(c_4) = \exp(2\pi i/5) + \exp(8\pi i/5),$$
$$\chi_V(c_5) = \exp(4\pi i/5) + \exp(6\pi i/5)$$

if $j = 1$ or 4, while

$$\chi_V(c_4) = \exp(4\pi i/5) + \exp(6\pi i/5),$$
$$\chi_V(c_5) = \exp(2\pi i/5) + \exp(8\pi i/5)$$

if $j = 2$ or 3.

In any case, one has that τ does not appear in V (either by Frobenius reciprocity or by calculating $\langle \chi_V, \chi_1 \rangle = 0$) and α_4 and α_5 each appear in V with multiplicity 1 (by calculating $\langle \chi_V, \chi_4 \rangle = \langle \chi_V, \chi_5 \rangle = 1$), so their complement is an irreducible representation of degree 3 (which checks with $\langle \chi_V, \chi_V \rangle = 3$), whose character is $\chi_V - \chi_4 - \chi_5$.

Choose $j = 1$ (or 4) and denote this representation by α_3, and choose $j = 2$ (or 3) and denote this representation by α_3' (and note that they are distinct as their characters are unequal). Then we may calculate that

$$\chi_3(c_1) = \chi_3'(c_1) = 3,$$
$$\chi_3(c_2) = \chi_3'(c_2) = -1,$$
$$\chi_3(c_3) = \chi_3'(c_3) = 0,$$
$$\chi_3(c_4) = 1 + \exp(2\pi i/5) + \exp(8\pi i/5) = (1 + \sqrt{5})/2,$$
$$\chi_3(c_5) = 1 + \exp(4\pi i/5) + \exp(6\pi i/5) = (1 - \sqrt{5})/2,$$

and vice versa for χ_3', agreeing with Example 3.5.26.

The definition of induction we have given is not the original one given by Frobenius. We now give his and show it is equivalent to ours when $[G : H]$ is finite. Actually, we give three definitions (although the second and third are almost the same). The first is Frobenius's.

Definition 1.27. Let G be a group and let H be a subgroup of G. Let $\sigma : H \to \mathrm{Aut}(W)$ be a representation of H.

(1) **(Frobenius)** Let $V' = \{f' : G \to W \mid f'(xh) = \sigma(h)^{-1} f'(x)$ for all $x \in G$, $h \in H\}$ and let $\rho' : G \to \mathrm{Aut}(V')$ be defined by

$$(\rho'(g)f')(x) = f'(g^{-1}x) \quad \text{for } g \in G, \ x \in G.$$

For (2) and (3), choose a system of left coset representatives $\{g_i\}$ of H with $g_1 = 1$.

(2) Let $V'' = \{f'' : G \to W \mid f''$ is constant on left H cosets$\}$ and let $\rho'' : G \to \mathrm{Aut}(V'')$ be defined by

$$(\rho''(g)f'')(x) = \sigma(g_i^{-1}gg_j)f''(g^{-1}x) \quad \text{where } x \in g_iH \text{ and } g^{-1}x \in g_jH.$$

(3) Let $V''' = \{f''' : G/H \to W\}$ and let $\rho''' : G \to \mathrm{Aut}(V''')$ be defined by

$$(\rho'''(g)f''')(g_iH) = \sigma(g_i^{-1}gg_j)f'''(g_jH) \quad \text{where } g_i^{-1}gg_j \in H.$$

Theorem 1.28.

(1) V' *is a representation of* G.

(2) V'' *is a representation of* G *that is isomorphic to* V'.

(3) V''' *is a representation of* G *that is isomorphic to* V''.

(4) *If* H *is a subgroup of* G *of finite index, then* V', V'', *and* V''' *are all isomorphic to* $V = \mathrm{Ind}_H^G(W)$.

Proof.

(1) We compute

$$(\rho'(g_1)(\rho'(g_2)f'))(x) = (\rho'(g_2)f')(g_1^{-1}x)$$
$$= f'(g_2^{-1}g_1^{-1}x)$$
$$= f'((g_1g_2)^{-1}x) = (\rho'(g_1g_2)f')(x)$$

as required.

(2) We begin by defining inverse isomorphisms $\alpha : V' \to V''$ and $\beta : V'' \to V'$.

First define $\alpha : V' \to V''$ by $\alpha(f') = f''$ where

$$f''(x) = \sigma(h)f'(x) \quad \text{where } x = g_ih.$$

Note that $f'' \in V''$ as if y is in the same left H-coset as x, then $y = x\bar{h} = g_ih\bar{h}$ for some $\bar{h} \in H$, and then

$$f''(y) = \sigma(h\bar{h})f'(y) = \sigma(h\bar{h})\sigma(\bar{h})^{-1}f'(x)$$
$$= \sigma(h)f'(x) = f''(x).$$

Next define $\beta : V'' \to V'$ by $\beta(f'') = f'$ where

$$f'(x) = \sigma(h)^{-1}f''(x) \quad \text{where } x = g_ih.$$

Note that $f' \in V'$ as if $\bar{h} \in H$,

$$f'(x\bar{h}) = \sigma(h\bar{h})^{-1}f''(x)$$
$$= \sigma(\bar{h})^{-1}\sigma(h)^{-1}f''(x) = \sigma(\bar{h})^{-1}f'(x).$$

Then, for any $x \in G$, writing $x = g_ih$,

$$(\beta\alpha(f'))(x) = \beta(f'')(x) \quad \text{where } f''(x) = \sigma(h)f'(x)$$
$$= \sigma(h)^{-1}f''(x) = \sigma(h)^{-1}\sigma(h)f'(x) = f'(x)$$

and

$$(\alpha\beta(f''))(x) = \alpha(f')(x) \quad \text{where } f'(x) = \sigma(h)^{-1}f''(x)$$
$$= \sigma(h)f'(x) = \sigma(h)\sigma(h)^{-1}f''(x) = f''(x)$$

so α and β are inverse isomorphisms.

We now use the following general principle: Let $\rho_1 : G \to \mathrm{Aut}(U_1)$ be a representation of G and let $\varphi : U_1 \to U_2$ be a vector space isomorphism. Then $\rho_2 : G \to \mathrm{Aut}(U_2)$ defined by

$$\rho_2(g)(u) = \varphi(\rho_1(g)(\varphi^{-1}(u))), \quad u \in U_2,$$

is a representation of G that is isomorphic to ρ_1. (This is simply the observation that if ρ_2 is defined in this way, then $\rho_2(g)(\varphi(u)) = \varphi(\rho_1(g)(u))$ for every $g \in G$ and $u \in U_1$.)

We apply this here with $U_1 = V'$, $U_2 = V''$, $\rho_1 = \rho'$, and $\varphi = \alpha$ (and hence $\varphi^{-1} = \beta$) to obtain $\rho'' = \rho_2$. It remains merely to explicitly calculate $\rho'' = \rho_2$.

We have, for $x \in G$, where we again set $x = g_i h$,

$$(\rho''(g)f'')(x) = \alpha(\rho'(g)(\beta(f'')))(x)$$
$$= \sigma(h)\rho'(g)(\beta(f''))(x)$$
$$= \sigma(h)\beta(f'')(g^{-1}x)$$

Now $g^{-1}x = g_j\overline{h}$ for some g_j and some $\overline{h} \in H$

$$= \sigma(h)\beta(f'')(g_j\overline{h})$$
$$= \sigma(h)\sigma(\overline{h})^{-1}f''(g_j\overline{h})$$
$$= \sigma(h\overline{h}^{-1})f''(g_j\overline{h})$$

But then $g_j\overline{h} = g^{-1}x = g^{-1}g_i h$ gives $h\overline{h}^{-1} = g_i^{-1}gg_j$

$$= \sigma(g_i^{-1}gg_j)f''(g^{-1}x)$$

as claimed.

(3) There is an obvious identification of V'' with V''', and, noting that, in the above notation, $g_i^{-1}gg_j = h\overline{h}^{-1} \in H$, under this identification ρ'' is identified with ρ'''.

(4) We prove this by showing that V'' is isomorphic to V. Let W_i be the subspace of V'' defined by

$$W_i = \{f'' \in V'' \mid f''(x) = 0 \text{ for } x \notin g_i H\}.$$

Then $V'' = \prod_i W_i = \bigoplus_i W_i$ as we are assuming $[G : H]$ finite. From the formula for $\rho(g'')f''$ we see that $\{W_i\}$ are permuted transitively under the action of G, and that

$$H = \{g \in G \mid \rho(g'')(W_1) = W_i\},$$

so by Theorem 1.6 we have that $V'' = \operatorname{Ind}_H^G(W_1)$. It remains to show that W_1 is isomorphic to W as a representation of H.

Let $\varepsilon : W_1 \to W$ be given by

$$\varepsilon(f'') = f''(1).$$

Then ε is an isomorphism of vector spaces, and, furthermore, for any $h \in H$,

$$\varepsilon(\rho''(h)f'') = (\rho''(h)f'')(1) = \sigma(h)f''(1) = \sigma(h)\varepsilon(f''),$$

so ε is an isomorphism of representations, completing the proof. $\qquad\square$

4.2. Mackey's Theorem

Our main result in this section is Mackey's theorem, which will generalize Corollary 1.10, but, more importantly, give a criterion for an induced representation to be irreducible. We begin with a pair of subgroups K, H of G. A K-H double coset is

$$KgH = \{kgh \mid k \in K, \quad h \in H\}.$$

It is easy to check that the K-H double cosets partition G (though, unlike ordinary cosets, they need not all have the same cardinality).

We shall also refine our previous notation slightly. Let $\sigma : H \to \operatorname{Aut}(W)$ be a representation of H. For $g \in G$, we shall set $H^g = gHg^{-1}$, and we will let σ^g be the representation $\sigma^g : H^g \to \operatorname{Aut}(W)$ by $\sigma^g(h) = \sigma(g^{-1}hg)$ for $h \in H^g$. Further, let us set $H_g = H^g \cap K$. We regard any representation of H, given by $\sigma : H \to \operatorname{Aut}(W)$, as a representation of H^g by σ^g and, hence, as a representation of H_g by the restriction of σ^g to H_g. We denote this representation of H^g (or H_g) by W_g. (In particular, this applies to the regular representation $\mathbf{F}(H)$ of H.) *Once again, $\mathbf{F}$, G, H, and K are completely arbitrary in this section unless stated otherwise.*

Theorem 2.1 (Mackey). *As $(\mathbf{F}(K), \mathbf{F}(H))$-bimodules,*

$$\mathbf{F}(G) \cong \bigoplus_g \mathbf{F}(K) \otimes_{\mathbf{F}(H_g)} \mathbf{F}(H),$$

where the sum is taken over a complete set of K-H double coset representatives.

Proof. For simplicity, let us write the right-hand side as $\bigoplus_g (\mathbf{F}(K) \otimes \mathbf{F}(H))_g$. Define maps α and $\widetilde{\beta}$ as follows.

For $g' \in G$, write g' as $g' = kgh$ for $k \in K$, $h \in H$, and g one of the given double coset representatives, and let $\alpha(g') = (k \otimes h)_g$. We must check that α is well defined. Suppose that $g' = \overline{k}g\overline{h}$ with $\overline{k} \in K$ and $\overline{h} \in H$. We need to show $(\overline{k} \otimes \overline{h})_g = (k \otimes h)_g$. Now $kgh = g' = \overline{k}g\overline{h}$ gives $g^{-1}k^{-1}\overline{k}g = h\overline{h}^{-1}$, and then

$$
\begin{aligned}
(\overline{k} \otimes \overline{h})_g &= (k(k^{-1}\overline{k}) \otimes (\overline{h}h^{-1})h)_g \\
&= (k \otimes g^{-1}(k^{-1}\overline{k})g(\overline{h}h^{-1})h)_g \\
&= (k \otimes h\overline{h}^{-1}(\overline{h}h^{-1})h)_g \\
&= (k \otimes h)_g
\end{aligned}
$$

as required. Then α extends to a map on $\mathbf{F}(G)$ by linearity. Conversely, define β_g on $K \times H$ by $\beta_g(k, h) = kgh \in G$ and extend β_g to a map

$$
\beta_g : \mathbf{F}(K) \times \mathbf{F}(H) \to \mathbf{F}(G)
$$

by linearity. Then for any $x \in H_g$, we have

$$
\beta_g(kx, h) = kxgh = kg(g^{-1}xg)h = \beta_g(k, g^{-1}xgh),
$$

so β_g is $\mathbf{F}(H_g)$-middle linear and so defines

$$
\widetilde{\beta}_g : (\mathbf{F}(K) \otimes \mathbf{F}(H))_g \to \mathbf{F}(G).
$$

Set $\widetilde{\beta} = \prod \widetilde{\beta}_g$. Then it is easy to check that α and $\widetilde{\beta}$ are inverses of each other, yielding the theorem. $\qquad\square$

Note that the subgroup H_g depends not only on the double coset KgH, but on the choice of representative g. However, the modules involved in the statement of Mackey's theorem are independent of this choice. We continue to use the notation of the preceding proof.

Proposition 2.2. *Let g and $\overline{g}$ be in the same K-H double coset of G. Then $(\mathbf{F}(K) \otimes \mathbf{F}(H))_g \cong (\mathbf{F}(K) \otimes \mathbf{F}(H))_{\overline{g}}$ as $(\mathbf{F}(K), \mathbf{F}(H))$-bimodules.*

Proof. In the notation of the previous proof, we have that

$$
\widetilde{\beta}_g : (\mathbf{F}(K) \otimes \mathbf{F}(H))_g \to \mathbf{F}(G)
$$

is an isomorphism of K-H bimodules onto its image.

Similarly we may construct

$$
\widetilde{\beta}_{\overline{g}} : (\mathbf{F}(K) \otimes \mathbf{F}(H))_{\overline{g}} \to \mathbf{F}(G),
$$

an isomorphism of K-H bimodules onto its image.

But $\mathrm{Im}(\widetilde{\beta}_g) = \mathbf{F}(KgH) = \mathbf{F}(K\overline{g}H) = \mathrm{Im}(\widetilde{\beta}_{\overline{g}})$. $\qquad\square$

Corollary 2.3. *For any* **F**-*representation* W *of* H,

$$\operatorname{Res}_K^G \operatorname{Ind}_H^G(W) = \bigoplus_g \left(\mathbf{F}(K) \otimes_{\mathbf{F}(H_g)} W \right) = \bigoplus_g \operatorname{Ind}_{H_g}^K(W_g).$$

Proof. By Theorem 2.1 and associativity of the tensor product, we have the following equalities among $\mathbf{F}(K)$-modules:

$$\operatorname{Ind}_H^G(W) = \mathbf{F}(G) \otimes_{\mathbf{F}(H)} W$$
$$= \left(\bigoplus_g \mathbf{F}(K) \otimes_{\mathbf{F}(H_g)} \mathbf{F}(H) \right) \otimes_{\mathbf{F}(H)} W$$
$$= \bigoplus_g \mathbf{F}(K) \otimes_{\mathbf{F}(H_g)} \left(\mathbf{F}(H) \otimes_{\mathbf{F}(H)} W \right)$$
$$= \bigoplus_g \mathbf{F}(K) \otimes_{\mathbf{F}(H_g)} W.$$

$\square$

Corollary 2.4. *Let* H *and* K *be arbitrary subgroups of* G. *Then*

$$\langle \operatorname{Ind}_K^G(\tau), \operatorname{Ind}_H^G(\tau) \rangle \;=\; \text{the number of } K\text{-}H \text{ double cosets in } G.$$

Proof. By Frobenius reciprocity and Corollary 2.3,

$$\langle \operatorname{Ind}_K^G(\tau), \operatorname{Ind}_H^G(\tau) \rangle \;=\; \langle \tau, \operatorname{Res}_K^G \operatorname{Ind}_H^G(\tau) \rangle$$
$$=\; \langle \tau, \bigoplus_g \operatorname{Ind}_{H_g}^K(\tau) \rangle$$
$$=\; \bigoplus_g \langle \operatorname{Res}_{H_g}^K(\tau), \tau \rangle$$
$$=\; |\{g\}|.$$

$\square$

Corollary 2.5. *Let* $K \subseteq H \subseteq G$ *with* K *a normal subgroup of* G. *Then*

$$\operatorname{Res}_K^G \operatorname{Ind}_H^G(W) = \bigoplus_g W_g,$$

where the sum is taken over a set of left coset representatives for H.

Proof. $\square$

Corollary 2.6. *Let* $G = KH$ *with* $K \cap H = \{1\}$. *Then*

$$\operatorname{Res}_K^G \operatorname{Ind}_H^G(W) = \dim(W)\mathbf{F}(K).$$

Proof. $\square$

Remark. If $G = KH$, $K \cap H = \{1\}$, with K or H normal, then G is a semidirect product. However, this need not be the case. For example, $S_m = S_{m-1}(\mathbb{Z}/m\mathbb{Z})$ with $S_{m-1} \cap (\mathbb{Z}/m\mathbb{Z}) = \{1\}$ but with neither subgroup normal for $m > 3$.

Lemma 2.7. *Let G be a finite group, H a subgroup of G, and $\mathbf{F}$ a field of characteristic zero or relatively prime to the order of H. Let W be an $\mathbf{F}$-representation of H defined by $\sigma : H \to \mathrm{Aut}(W)$ and set $V = \mathrm{Ind}_H^G(W)$. Then*

$$\mathrm{End}_G(V) \cong \bigoplus_g \mathrm{Hom}_{H_g}(W_g, \, \mathrm{Res}_{H_g}^H(W)),$$

where the direct sum is over a complete set of H-H double coset representatives, $H_g = H^g \cap H$, and W_g is the representation of H_g on W defined by σ^g.

Proof. The required isomorphism is a consequence of the following chain of equalities and isomorphisms.

$$\mathrm{End}_G(V) = \mathrm{Hom}_G(V, \, V)$$
$$\cong \mathrm{Hom}_H(W, \, V)$$

as in the proof of Frobenius reciprocity

$$\cong \mathrm{Hom}_H(V, \, W)$$

as by our assumption on $\mathbf{F}$, $\mathbf{F}(H)$ is semisimple

$$= \mathrm{Hom}_{\mathbf{F}(H)}\left(\bigoplus_g \mathbf{F}(H) \otimes_{\mathbf{F}(H_g)} W, \, W\right)$$
$$\cong \prod_g \mathrm{Hom}_{\mathbf{F}(H)}\left(\mathbf{F}(H) \otimes_{\mathbf{F}(H_g)} W, \, W\right).$$

Now note that in the first W above, H_g is operating by σ^g; while in the second, it is operating by σ

$$\cong \bigoplus_g \mathrm{Hom}_{\mathbf{F}(H_g)}\left(W_g, \, \mathrm{Res}_{H_g}^H(W)\right)$$

by adjoint associativity of Hom and tensor product (Theorem A.1.4). (The direct sum and direct product agree here as $\{g\}$ is finite.) $\qquad\square$

Corollary 2.8.

(1) *Let G be an arbitrary group, H a finite subgroup of G, and $\mathbf{F}$ a field of characteristic zero or relatively prime to the order of H. Let W be a representation of H and set $V = \operatorname{Ind}_H^G(W)$. Then $\operatorname{End}_G(V) = \mathbf{F}$ if and only if $\operatorname{End}_H(W) = \mathbf{F}$ and $\operatorname{Hom}_{H_g}(W_g,\ \operatorname{Res}_{H_g}^H(W)) = 0$ for every $g \in G$, $g \notin H$.*

(2) *Let G be a finite group. Let $\mathbf{F}$ be a field of characteristic zero or relatively prime to the order of G. Then V is absolutely irreducible if and only if W is absolutely irreducible and, for each $g \in G$, $g \notin H$, the H_g-representations W_g and $\operatorname{Res}_{H_g}^H(W)$ are disjoint (i.e., have no mutually isomorphic irreducible components). In particular, if H is a normal subgroup of G, V is absolutely irreducible if and only if W is absolutely irreducible and distinct from all its conjugates.*

Proof. $\operatorname{End}_G(V)$ contains a subspace isomorphic to $\operatorname{End}_H(W)$ (given by the double coset representative $g = 1$). Also, under our assumptions V is absolutely irreducible if and only if $\operatorname{End}_G(V) = \mathbf{F}$, and similarly for W. $\square$

Example 2.9. Let $H = \mathbb{Z}/m\mathbb{Z}$ and $G = D_{2m}$. Then the representations θ_i and θ_{m-i} of H are conjugate, so for $i \neq m/2$, $\operatorname{Ind}_H^G(\theta_i)$ is irreducible. Of course, this representation is just φ_i.

Example 2.10. Let $H = A_5$ and $G = S_5$. As a system of coset representatives, we choose $\{1,\ (1\ 2\ 5\ 4)\} = \{g_1,\ g_2\}$. Note that

$$g_2(1\ 2\ 3\ 4\ 5)g_2^{-1} = (1\ 3\ 5\ 2\ 4) = (1\ 2\ 3\ 4\ 5)^2.$$

Thus, if we let σ^1 be the representation α_3 of A_5 (in the notation of Example 3.5.26), we see that its conjugate $\sigma^2 = \alpha_3'$, a distinct irreducible representation. Hence by Corollary 2.8, $\operatorname{Ind}_H^G(\alpha_3) = \operatorname{Ind}_H^G(\alpha_3')$ is an irreducible representation of degree 6, and by Theorem 1.20 we may compute its character, giving an alternative to the method of Example 3.5.27.

Example 2.11. Let $H = V$ and $G = A_4$ in the notation of Example 1.18. Then the representations λ_1, λ_2, and λ_3 of H are mutually conjugate, so $\operatorname{Ind}_H^G(\lambda_i)$ is irreducible and is equal to α for $i = 1$, 2, 3, verifying the result of Example 1.18.

Example 2.12. Let m be such that $p = 2^m - 1$ is prime. Then there is an action of $S = \mathbb{Z}/p\mathbb{Z}$ on $V = (\mathbb{Z}/2\mathbb{Z})^m$ by linear transformations that cyclically permutes the elements of V other than the identity. (This may most easily be seen by observing that $GL(m,\ \mathbf{F}_2)$ has an element of order p.) Thus, we may form the semidirect product

$$1 \longrightarrow V \longrightarrow G \overset{\pi}{\longrightarrow} S \longrightarrow 1$$

with G a group of order $n = 2^m(2^m - 1)$. G has the 1-dimensional complex representations $\pi^*(\theta_i)$ for $i = 0, \ldots, p-1$. Also, for *any* nontrivial irreducible complex representation σ of V, G has the representation $\mathrm{Ind}_V^G(\sigma)$ of degree $[G : V] = 2^m - 1$. Now σ is disjoint from all its conjugates (as $\mathrm{Ker}(\sigma)$ may be considered to be an $\mathbf{F}_2$-vector space of dimension $m - 1$, and $GL(m-1, \mathbf{F}_2)$ does not have an element of order p), so by Corollary 2.8, σ is irreducible. As $(2^m - 1)^2 + (2^m - 1)(1)^2 = n$, these 2^m complex representations are all of the irreducible complex representations of G. (Note that if $m = 2$, then $G = A_4$, so this is a generalization of Example 2.10.)

Construction 2.13 (Mackey). Let the finite group G be the semidirect product of a subgroup H by an abelian normal subgroup A, i.e.,

$$1 \to A \to G \to H \to 1$$

with A normal in G. Let $\{\sigma\}$ be the set of irreducible (i.e., 1-dimensional) representations of A over an algebraically closed field $\mathbf{F}$ of characteristic 0 or prime to $|G|$. Then G acts on σ by $\sigma^g(a) = \sigma(g^{-1}ag)$. Observe that this action factors through H.

Let $\{\sigma_i\}$ be a system of representatives for the orbits of G on $\{\sigma\}$. For each i, let $H_i = \{h \in H \mid \sigma_i^h = \sigma_i\}$, and let G_i be the semidirect product $G_i = AH_i$. Note that σ_i extends to a representation of G_i by $\sigma_i(ah) = \sigma_i(a)$, $a \in A$, $h \in H_i$, as

$$\sigma_i(a_1 h_1 a_2 h_2) = \sigma_i(a_1 h_1 a_2 h_1^{-1} h_1 h_2) = \sigma_i(a_1 h_1 a_2 h_1^{-1})$$
$$= \sigma_i(a_1)\sigma_i(h_1 a_2 h_1^{-1}) = \sigma_i(a_1)\sigma_i(a_2) = \sigma_i(a_1 h_1)\sigma_i(a_2 h_2).$$

Thus σ_i is a 1-dimensional representation of G_i. Let ρ be an irreducible representation of H_i. Then $\sigma_i \otimes \pi^*(\rho)$ is an irreducible representation of G_i (where $\pi : G_i \to H_i$ is the projection.)

Define $\theta_{i,\rho}$ by

$$\theta_{i,\rho} = \mathrm{Ind}_{G_i}^G(\sigma_i \otimes \pi^*(\rho)).$$

Theorem 2.14.

(1) *Each $\theta_{i,\rho}$ is an irreducible representation of G.*

(2) *If $\theta_{i,\rho}$ and $\theta_{i',\rho'}$ are isomorphic, then $i = i'$ and $\rho = \rho'$.*

(3) *$\{\theta_{i,\rho}\}$ are all of the irreducible representations of G.*

Proof. (1) Note that, for any $g \in G$, $AgH_i = gAH_i = gG_i$, so a system of left G_i coset representatives serves as a system of A-H_i double coset representatives.

For simplicity of notation, let $L = G_i$ and $L_g = G_i \cap gG_ig^{-1}$ (so $L_1 = L$). Observe that $L_g \supseteq A$ for every $g \in G$, and also that if $W_1 = W = \sigma_i \otimes \pi^*(\rho)$, then $W_g = \mathrm{Res}^L_{L_g}(\sigma_i^g \otimes \pi^*(\rho))$.

Then, as $\pi^*(\rho)$ is trivial on A,

$$\mathrm{Res}^{L_g}_A(W_g) = d\sigma_i^g, \quad d = \dim(W).$$

But as g runs over a set of left G_i coset representatives, the $\{\sigma_i^g\}$ are pairwise mutually disjoint representations of A, so by Corollary 2.8, $\theta_{i,\rho}$ is irreducible.

(2) Note that

$$\begin{aligned}
\mathrm{Res}^G_A(\theta_{i,\rho}) &= \mathrm{Res}^G_A \mathrm{Ind}^G_{G_i}(\sigma_i \otimes \pi^*(\rho)) \\
&= \bigoplus_g (\sigma_i \otimes \pi^*(\rho))_g \\
&= \bigoplus_g d\sigma_i^g
\end{aligned}$$

consists entirely of multiples of representations of A in the G-orbit of σ_i, so $\theta_{i,\rho}$ determines i.

Let $\{g_j\}_{j \in J}$ be a system of left coset representatives of G_i and set $W_j = W_{g_j}$. Then if $V = \theta_{i,\rho}$, we have, in the notation of Theorem 1.6, that $V = \bigoplus_{j \in J} W_j$.

If $g = g_j$, then on W_j we have

$$\theta_{i,\rho}(a)(w) = \sigma_i^g(a)(w), \quad w \in W_j.$$

Thus, in particular,

$$W_1 = \{v \in V \mid \theta_{i,\rho}(a)(v) = \sigma_i(a)(v) \text{ for all } a \in A\}.$$

But on W_1,

$$(\sigma_i \otimes \pi^*(\rho))(h)(w) = \rho(h)(w) \quad \text{for every } h \in H_i, \ w \in W.$$

Thus, on the nonempty subspace

$$\{v \in V \mid \theta_{i,\rho}(a)(v) = \sigma_i(a)(v) \text{ for all } a \in A\},$$

the representation of H_i is ρ, so $\theta_{i,\rho}$ determines ρ.

(3) $\{\sigma\}$ has cardinality $|A|$ and the orbit of σ_i has cardinality $|H|/|H_i|$, so we see

$$|A| = \sum_i |H|/|H_i|.$$

For fixed i, the sum of the squares of the degrees of the representations ρ is $|H_i|$, so the sum of the squares of the degrees of the representations $\theta_{i,\rho}$ is

$$[G : G_i]^2 |H_i| = [H : H_i]^2 |H_i| = |H|^2/|H_i|$$

and hence the sum of the squares of the degrees of all the $\theta_{i,\rho}$ is

$$\sum_i |H|^2/|H_i| = |H| \sum_i |H|/|H_i| = |H||A| = |G|,$$

so these are all the irreducible representations.

Remark. Our exposition here follows Serre [Se, Section 8.2].

Example 2.15. Let $H \subseteq \mathbf{F}_p^*$ be a subgroup of the multiplicative group of the finite field $\mathbf{F}_p$, and let A be the additive group of $\mathbf{F}_p$. Then

$$G = \left\{ \begin{bmatrix} h & a \\ 0 & 1 \end{bmatrix} \;\middle|\; \begin{array}{l} a \in \mathbf{F}_p \\ h \in H \end{array} \right\}$$

is a semidirect product

$$1 \to A \to G \xrightarrow{\pi} H \to 1,$$

where we identify A with $\left\{ \begin{bmatrix} 1 & a \\ 0 & 1 \end{bmatrix} \right\}$ and H with $\left\{ \begin{bmatrix} h & 0 \\ 0 & 1 \end{bmatrix} \right\}$. Let $q = |H|$, so $|G| = pq$.

A has complex representations $\tau = \theta_0^p, \theta_1^p, \ldots, \theta_{p-1}^p$ (where in θ_i^p, $\begin{bmatrix} 1 & 1 \\ 0 & 1 \end{bmatrix}$ acts by multiplication by ζ^i, where $\zeta = \exp(2\pi\sqrt{-1}/p)$.) Note that

$$\begin{bmatrix} h^{-1} & 0 \\ 0 & 1 \end{bmatrix} \begin{bmatrix} 1 & a \\ 0 & 1 \end{bmatrix} \begin{bmatrix} h & 0 \\ 0 & 1 \end{bmatrix} = \begin{bmatrix} 1 & ah^{-1} \\ 0 & 1 \end{bmatrix}.$$

From this we see that if $\sigma = \theta_i^p$, then if $i = 0$, $H_i = H$, while if $i \neq 0$, $H_i = \{1\}$.

First consider $i = 0$, so $\sigma_i = \theta_0^p = \tau$. Then $H = H_0$ has complex representations $\tau = \theta_0^q, \ldots, \theta_{q-1}^q$. Also, $G_i = AH_0 = AH = G$, so we obtain the distinct irreducible 1-dimensional representations $\pi^*(\theta_i^q)$, $i = 0, \ldots, q-1$, of G.

Now consider $i \neq 0$. Then $H_i = \{1\}$ so $G = AH_i = A$ and we obtain the distinct irreducible q-dimensional representations $\mathrm{Ind}_A^G(\theta_i^p)$, where i varies over a system of coset representatives of $\mathbf{F}_p^*/H$ (a set whose cardinality is $(p-1)/q$).

Thus we have obtained all irreducible complex representations of G.

We may verify this construction by character computations. We certainly have the q distinct irreducible 1-dimensional representations $\pi^*(\theta_i^q)$, $i = 0, \ldots, q - 1$.

Let $V_i = \mathrm{Ind}_A^G(\theta_i^p)$, $i \neq 0$, and let $\zeta = \exp(2\pi\sqrt{-1}/p)$. From Theorem 1.20, it is easy to check that if $\chi_i = \chi_{V_i}$,

$$\chi_i\left(\begin{bmatrix} x & a \\ 0 & 1 \end{bmatrix}\right) = \begin{cases} 0 & \text{if } x \neq 1, \\ \displaystyle\sum_{h \in H} \zeta^{ih^{-1}a} & \text{if } x = 1. \end{cases}$$

Then

$$\langle \chi_i, \chi_j \rangle = \frac{1}{pq} \sum_{g \in G} \chi_i(g)\chi_j(g^{-1})$$

$$= \frac{1}{pq} \sum_{\substack{h_1 \in H \\ h_2 \in H}} \sum_{a \in A} (\zeta^{ih_1^{-1} - jh_2^{-1}})^a$$

$$= \begin{cases} 0 & \text{if } ih_i^{-1} - jh_2^{-1} \text{ is never } 0 \ (\mathrm{mod}\ p), \\ 1 & \text{if } ih_1^{-1} - jh_2^{-1} \text{ is sometimes } 0 \ (\mathrm{mod}\ p) \end{cases}$$

(as in the latter case, $ih_1^{-1} - jh_2^{-1} \equiv 0 \ (\mathrm{mod}\ p)$ q times, and the sum is p each time that occurs).

But $ih_1^{-1} - jh_2^{-1} \equiv 0 \ (\mathrm{mod}\ p)$ is equivalent to $i \equiv jh_2^{-1}h_1 \ (\mathrm{mod}\ p)$, which occurs if and only if i and j are in the same coset of H in $\mathbf{F}_p^*$.

Thus $\langle \chi_i, \chi_i \rangle = 1$ so χ_i is irreducible, and $\{\chi_i\}$ are distinct as i ranges over a set of coset representatives of H in $\mathbf{F}_p^*$.

Finally,

$$q(1)^2 + ((p - 1)/q)(q^2) = pq = |G|,$$

so these are all the irreducible complex representations of G.

4.3. Permutation Representations

We have already encountered permutation representations, but because of their particular importance, we wish to investigate them further. We shall restrict our attention to complex representations, so that we may use the full power of character theory. We begin with a bit of recapitulation.

Definition 3.1. Let $P = \{p_i\}_{i \in I}$ be a set and $\sigma : G \to \mathrm{Aut}(P)$ a homomorphism. Then σ defines a **permutation representation** on $\mathbf{C}P$ (the complex vector space with basis P) by

$$\left(\sum_{g \in G} a_g g\right)\left(\sum_{i \in I} a_i p_i\right) = \sum_{g,\,i} a_g a_i \sigma(g)(p_i).$$

We will often call this representation σ as well.

For simplicity, we shall assume throughout that G and P are both finite, though some of our results hold more generally.

Lemma 3.2. *Let σ be a permutation representation of G on P. Then for any $g \in G$,*

$$\chi_\sigma(g) = |\{p \in P \mid \sigma(g)(p) = p\}|.$$

Proof. Consider the matrix $[\sigma(g)]$ of $\sigma(g)$ in the basis P of $\mathbf{C}P$. This is a permutation matrix, with an entry of 1 on the diagonal for each $p \in P$ with $g(p) = p$, and entries of 0 on the diagonal otherwise. Thus the trace of this matrix, and hence the value of $\chi_\sigma(g)$, is equal to the number of elements of P that are fixed by g. $\qquad\square$

The following terminology is standard. Let $p \in P$. Then the **orbit** $\mathcal{O}_p$ of p and the **stabilizer** G_p of p are defined by

$$\mathcal{O}_p = \{p' \in P \mid \sigma(g)(p) = p' \quad \text{for some } g \in G\},$$
$$G_p = \{h \in G \mid \sigma(h)(p) = p\}.$$

Note that $\mathcal{O}_p$ is a subset of P and G_p is a subgroup of G. Observe also that $|\mathcal{O}_p||G_p| = |G|$.

Definition 3.3. A nonempty subset Q of P is acted on **transitively** by σ if $\sigma(g)(q) \in Q$ for every $g \in G$ and $q \in Q$, and furthermore if for every $q_1, q_2 \in \theta$ there is a $g \in G$ with $\sigma(g)(q_1) = q_2$. In this situation we call Q a **domain of transitivity**. If P is a domain of transitivity, then σ is called **transitive**.

Note that the domains of transitivity are just the distinct orbits.

The following is obvious:

Lemma 3.4. *Let $Q_1, \ldots, Q_k$ partition P into domains of transitivity. Then*

$$\mathbf{C}P = \mathbf{C}Q_1 \oplus \cdots \oplus \mathbf{C}Q_k.$$

Proof. $\qquad\square$

Theorem 3.5. *Let σ be a transitive permutation representation of G on P and let $p \in P$. Then*

$$\sigma = \mathrm{Ind}_{G_p}^G(\tau),$$

and for any $g \in G$,

$$\chi_\sigma(g) = |P||C_g \cap G_p|/|C_g|,$$

where C_g denotes the conjugacy class of G containing g.

Proof. The first statement is Corollary 1.8 (1) and the second is Corollary 1.22 (3), observing that $[G : G_p] = |P|$. $\qquad\qquad\qquad\qquad\qquad\square$

Observe that in this situation all of the subgroups G_p, for $p \in P$, are mutually conjugate so we may choose p arbitrarily (compare Corollary 1.7).

Example 3.6. We shall verify Theorem 3.5 in a couple of concrete cases.

(1) Let S_m be the symmetric group on m symbols, and consider its natural representation σ on the set $P = \{1,\dots,m\}$. Let $g \in S_m$ be a transposition. We may calculate $\chi_\sigma(g)$ in two ways. The easy way is to use Lemma 3.2: g fixes $m - 2$ elements of P so $\chi_\sigma(g) = m - 2$. The hard way is to use Theorem 3.5: Let $p \in P$. Then G_p is isomorphic to the symmetric group on $m - 1$ symbols. The conjugacy class of an element of a symmetric group is determined by its cycle structure, so $|C_g|$ is the number of transpositions in S_m, which is $\binom{m}{2}$, and $|C_g \cap G_p|$ is the number of transpositions in S_{m-1}, which is $\binom{m-1}{2}$. Then $\chi_\sigma(g) = m\binom{m-1}{2}/\binom{m}{2} = m - 2$.

(2) Let S_m be the symmetric group on m symbols, and consider its natural representation σ on the set P of unordered k-tuples of elements of $\{1,\dots,m\}$. Observe $|P| = \binom{m}{k}$. Let $g \in S_m$ be a j-cycle. We consider the most interesting case, where $j \leq k$ and $j \leq m - k$. Again we compute $\chi_\sigma(g)$ in two ways. This time we use Theorem 3.5 first: Let $p \in P$. Then G_p is isomorphic to $S_k \times S_{m-k}$. Again, the conjugacy class of an element of a symmetric group is determined by its cycle structure, so $|C_g|$ is the number of j-cycles in S_m, which is $\frac{m!}{(m-j)!j}$ (to obtain a j-cycle, choose j elements of $\{1,\dots,m\}$ in order; there are $m \cdots (m - j + 1) = \frac{m!}{(m-j)!}$ ways to do this, and this procedure yields every j-cycle exactly j times), and $|C_g \cap G_p|$ is the number of j-cycles in $S_k \times S_{m-k}$, which by the same logic is $\frac{k!}{(k-j)!j} + \frac{(m-k)!}{(m-k-j)!j}$. Then

$$\chi_\sigma(g) = \frac{\binom{m}{k}\left(\frac{k!}{(k-j)!j} + \frac{(m-k)!}{(m-k-j)!j}\right)}{\frac{m!}{(m-j)!j}}.$$

Now we use Lemma 3.2: A j-cycle g fixes a k-tuple p if and only if either p contains the j elements of $\{1,\dots,m\}$ "moved" by g or the complement of p in $\{1,\dots,m\}$ contains the j elements of $\{1,\dots,m\}$ moved by g. To extend a j-element subset of $\{1,\dots,m\}$ to an i-element subset of $\{1,\dots,m\}$ we must choose $i - j$ of the remaining $m - j$ elements, so there are $\binom{m-j}{i-j}$ ways to do this; letting $i = k$ or $i = m - k$ we see that

$$\chi_\sigma(g) = \binom{m-j}{k-j} + \binom{m-j}{m-k-j},$$

and these two numbers are equal.

Because of Lemma 3.4, we shall almost always restrict our attention to transitive representations, though we state the next three results more generally.

Lemma 3.7. *Let σ_i be a permutation representation of G on a set P_i for $i = 1, 2$. Then σ_1 and σ_2 are equivalent if and only if for each $g \in G$,*

$$|\{p_1 \in P_1 \mid \sigma_1(g)(p_1)\}| = |\{p_2 \in P_2 \mid \sigma_2(g)(p_2) = p_2\}|.$$

Proof. We know that σ_1 and σ_2 are equivalent if and only if their characters χ_1 and χ_2 are equal. But, by Lemma 3.2, for each $g \in G$, the values of $\chi_1(g)$ and $\chi_2(g)$ are given by the left and right hand sides of the above equation. $\square$

Proposition 3.8. *Let P be partitioned into k domains of transitivity under the representation σ of G. Then the multiplicity of τ in $\mathbf{C}P$ is equal to k.*

Proof. By Lemma 3.4, we may assume $k = 1$. But then, by Theorem 3.5 and Frobenius reciprocity,

$$\langle \tau, \sigma \rangle \;=\; \langle \tau, \mathrm{Ind}_H^G(\tau) \rangle \;=\; \langle \mathrm{Res}_H^G(\tau), \tau \rangle \;=\; 1.$$

$\square$

Corollary 3.9 (Cauchy-Frobenius). *Let P be partitioned into k domains of transitivity under the representation σ of G. Then the average number of fixed points of elements of G on P is equal to k.*

Proof. By Proposition 3.8,

$$k \;=\; \langle \sigma, \tau \rangle \;=\; \frac{1}{n}\sum_{g \in G} \chi_\sigma(g)\chi_\tau(g) \;=\; \frac{1}{n}\sum_{g \in G}\chi_\sigma(g),$$

and, by Lemma 3.2, $\chi_\sigma(g)$ is equal to the number of elements of P fixed by g. $\square$

Corollary 3.10 (Burnside). *Let σ be a transitive permutation representation of G on P. Let $p \in P$ and set $H = G_p$. Then $\langle \sigma, \sigma \rangle$ is equal to the number of orbits of $\mathrm{Res}_H^G(\sigma)$ on P.*

Proof. In this case, by Corollary 2.4, $\langle \sigma, \sigma \rangle = \langle \operatorname{Ind}_H^G(\tau), \operatorname{Ind}_H^G(\tau) \rangle$ is equal to the number of H-H double cosets in G. But it is easy to check that there is a one-to-one correspondence between H-H double cosets and orbits of H on P. $\square$

Definition 3.11. A partition of P into subsets $\{Q_i\}_{i \in I}$ is called a partition into **domains of imprimitivity** for σ if for every $g \in G$ and every $i \in I$ there exists a $j \in I$ with $\sigma(g)(Q_i) = Q_j$. If the only partitions into domains of imprimitivity are either the partition consisting of the single set $Q = P$, or the partition into subsets of P consisting of single elements, then σ is called **primitive**, otherwise **imprimitive**.

(Note that an intransitive representation of G is certainly imprimitive.)

Remark 3.12. Let σ be a transitive permutation representation of G on P and let Q a subset of P with the property that for every $g \in G$, either $\sigma(g)(Q) = Q$ or $\sigma(g)(Q) \cap Q = \emptyset$. Set

$$H = \{g \in G \mid \sigma(g)(Q) = Q\}.$$

Then H is a subgroup of G, and if $\{g_i\}$ are a set of left coset representatives of H, then $Q_i = g_i(Q)$ partitions P into domains of imprimitivity. Furthermore, all partitions into domains of imprimitivity arise in this way. Note that H acts as a group of permutations on Q.

We have the following result, generalizing Theorem 3.5.

Theorem 3.13. *Let σ be a transitive permutation representation of G on P, Q a domain of imprimitivity for σ, and $H = \{g \in G \mid \sigma(g)(Q) = Q\}$. If ρ denotes the permutation representation of H on Q given by $\rho(h)(q) = \sigma(h)(q)$ for $h \in H$, $q \in Q$, then ρ is a transitive permutation representation of H on Q and*

$$\sigma = \operatorname{Ind}_H^G(\rho).$$

Proof. This follows directly from Theorem 1.6. $\square$

We have the following useful proposition.

Proposition 3.14. *Let σ be a transitive permutation representation of G on a set P, and let $p \in P$.*

(1) *Let P be partitioned into a set of domains of imprimitivity $\{Q_i\}$ with $p \in Q_1$, and let*

$$H = \{g \in G \mid \sigma(g)(Q_1) = Q_1\}.$$

Then there are $[G : H]$ sets in the partition, each of which has $[H : G_p]$ elements.

(2) *Let H be a subgroup of G containing G_p, and let*

$$Q = \{\sigma(g)(p) \mid g \in H\}.$$

Then Q is one element in a partition of P into $[G : H]$ domains of imprimitivity, each of which contains $[H : G_p]$ elements.

Proof. This is clear from Remark 3.12, identifying domains of imprimitivity with left cosets. $\square$

Corollary 3.15. *Let σ be a transitive permutation representation of G on a set P.*

(1) *σ is primitive if and only if some (and hence every) G_p is a maximal subgroup of G.*

(2) *If $|P|$ is prime, then σ is primitive.*

Proof. (1) is clear. As for (2), by Proposition 3.14, the cardinality of a domain of imprimitivity for G divides $[G : G_p] = |P|$, so if $|P|$ is prime, then a domain of imprimitivity consists of either a single element or all of P. $\square$

We now introduce another sort of property of a representation.

Definition 3.16. A permutation representation σ of a group G on a set P is **k-fold transitive** if P has at least k elements and for any pair $(p_1, \ldots, p_k)$ and $(q_1, \ldots, q_k)$ of k-tuples of distinct elements of P there is a $g \in G$ with $\sigma(g)(p_i) = q_i$ for $i = 1, \ldots, k$.

Examples 3.17.

(1) Note that 1-fold transitive is just transitive.

(2) The permutation representation of D_{2m} on the vertices of an m-gon is doubly transitive if $m = 3$, but only singly ($=$ 1-fold) transitive if $m > 3$.

(3) The natural permutation representation of S_m on $\{1, \ldots, m\}$ is m-fold transitive, and of A_m on $\{1, \ldots, m\}$ is $(m-2)$-fold (but not $(m-1)$-fold) transitive.

(4) The natural permutation representation of S_m on (ordered or unordered) pairs of elements of $\{1, \ldots, m\}$ is transitive but *not* doubly ($=$ 2-fold) transitive for $m > 3$, for there is no $g \in S_m$ taking $\{(1,2),(2,3)\}$ to $\{(1,2),(3,4)\}$.

Doubly transitive permutation representations have two useful properties:

Proposition 3.18. *Let σ be a doubly transitive permutation representation of G on a set P. Then σ is primitive.*

Proof. Suppose σ is not primitive, and let Q be a domain of imprimitivity, with $p_1, p_2 \in Q$ and $p_3 \notin Q$. Then there is no $g \in G$ with $\sigma(g)(p_1) = p_1$ and $\sigma(g)(p_2) = p_3$, so σ is not doubly transitive. The proposition follows by contraposition. $\qquad\square$

Theorem 3.19. *Let σ be a transitive permutation representation of G on a set P. Then σ is doubly transitive if and only if $\sigma = \tau \oplus \sigma'$ for some irreducible representation σ' of G.*

Proof. Since σ is a permutation representation, its character χ is real valued.

Note that χ^2 is the character of the permutation representation $\sigma \otimes \sigma$ on $P \times P$. Let k be the number of orbits of $\sigma \otimes \sigma$ on $P \times P$. Note that $k = 2$ if σ is doubly transitive (the orbits being $\{(p,p) \mid p \in P\}$ and $\{(p,q) \mid p, q \in P,\ p \neq q\}$) and $k > 2$ otherwise. Then, by Proposition 3.8,

$$
\begin{aligned}
k &= \langle \sigma \otimes \sigma, \tau \rangle \\
&= \frac{1}{n} \sum_{g \in G} \chi^2(g) \\
&= \frac{1}{n} \sum_{g \in G} \chi(g)\chi(g) \\
&= \frac{1}{n} \sum_{g \in G} \chi(g)\overline{\chi}(g) \\
&= \langle \chi, \chi \rangle.
\end{aligned}
$$

Note that τ appears in σ with multiplicity 1, again by Proposition 3.8. Note also that $\langle \chi, \chi \rangle = 2$ if and only if in the decomposition of σ into irreducibles there are exactly two distinct summands, yielding the theorem. $\qquad\square$

Example 3.20. The representation β of Example 3.5.26 was doubly transitive and decomposed as $\tau \oplus \alpha_4$, α_4 irreducible. The representation γ of Example 3.5.26 was not doubly transitive: $\gamma \otimes \gamma$ had three orbits on $\{1, \dots, 5\} \times \{1, \dots, 5\}$, and γ decomposed into a sum $\tau \oplus \alpha_4 \oplus \alpha_5$ of three distinct irreducibles.

Remark 3.21. The converse of Proposition 3.18 is false. For example, let p be a prime and consider the permutation representation of D_{2p} on the vertices of a p-gon. By Corollary 3.15 (2), this representation is primitive, but for $p > 3$ it is not doubly transitive.

The action of a group on itself by left multiplication is a permutation representation. Of course, this is nothing other than the regular representation, which we have already studied extensively. Instead, we consider the action of a group on itself by *conjugation*, i.e., $\gamma(g)(h) = ghg^{-1}$ for all

$g, h \in G$. Let us determine some examples of this representation. Of course, if G is abelian, this representation is just $n\tau$. In any case, the orbits of this representation are just the conjugacy classes. In general, if $C_1, \ldots, C_t$ are the conjugacy classes of G (in some order), we will let γ_i be γ on C_i, so

$$\gamma = \gamma_1 \oplus \cdots \oplus \gamma_t.$$

For an element g of G, we define its centralizer $C(g)$ to be the subgroup $C(g) = \{x \in g \mid xgx^{-1} = g\}$. Then if $c_i \in C_i$, $i = 1, \ldots, t$, and $H_i = C(g_i)$, we have that $\gamma_i = \mathrm{Ind}_{H_i}^{G}(\tau)$, for each i. Of course, $\gamma_1 = \tau$, and each γ_i contains τ once, as each γ_i is a transitive permutation representation, so τ appears in γ with multiplicity t. While γ is a natural representation to study, in general it is not well understood. However, we can easily compute its character.

Theorem 3.22 (Frame). *Let γ be the conjugation representation of G, and let χ_i, $i = 1, \ldots, t$, be the characters of the irreducible representations of G. Then for every $g \in G$,*

$$\chi_\gamma(g) = \sum_{i=1}^{t} \chi_i(g)\chi_i(g^{-1}).$$

Proof. Let g be in conjugacy class C. Then, by Corollary 3.5.18, the right-hand side is $n/|C|$. On the other hand, the left-hand side is

$$\begin{aligned}
\chi_\gamma(g) &= |\{x \in G \mid gxg^{-1} = x\}| \\
&= |\{x \in G \mid g = xgx^{-1}\}| \\
&= |C(g)| = n/|C|,
\end{aligned}$$

and these agree. $\qquad\qquad\square$

Example 3.23.

(1) Consider D_{2m} for m odd. Then (cf. Example 3.4.1) we have

$$C_1 = \{1\}, \quad C_2 = \{x, x^{m-1}\}, \ldots, \quad C_{(m+1)/2} = \{x^{(m-1)/2}, x^{(m+1)/2}\},$$
$$C_{(m+3)/2} = \{y, xy, \ldots, x^{m-1}y\}.$$

Then $\gamma_1 = \tau$, and for $i = 2, \ldots, (m+1)/2$, γ_i is a nontrivial representation containing τ as a subrepresentation. Since D_{2m} only has one nontrivial 1-dimensional representation, namely, ψ_-, we see that

$$\gamma_i = \tau \oplus \psi_- \qquad \text{for } 2 \le i \le (m+1)/2.$$

We are left with the action on $C_{(m+3)/2}$, and we claim this is $\mathbf{C}P$ (cf. Example 3.1.4 (7)), where $P = \{\text{vertices of a regular } m\text{-gon}\}$. This may be seen as follows: $C_{(m+3)/2}$ consists of the elements of order exactly two in D_{2m}, and each such fixes exactly one vertex of P. Thus we have a one-to-one correspondence

$$C_{(m+3)/2} \longleftrightarrow P$$

by

$$x^i y \longrightarrow \text{vertex } v \text{ fixed by } x^i y$$

and further, for any $g \in D_{2m}$, $g(x^i y)g^{-1}$ fixes the vertex $g(v)$, so the two actions of G are isomorphic.

Now it is easy to check directly that

$$\mathbf{C}P = \tau \oplus \varphi_1 \oplus \cdots \oplus \varphi_{(m-1)/2}.$$

Alternatively, $\mathbf{C}P = \mathrm{Ind}_H^G(\tau)$, where H is the subgroup fixing a single vertex v, so $H = \{1, x^i y\}$ for some i, and we may choose $H = \{1, y\}$. Then the calculation of $\mathbf{C}P$ is made even easier by Frobenius reciprocity: $\langle \sigma, \mathrm{Ind}_H^G(\tau) \rangle = \langle \mathrm{Res}_H^G(\sigma), \tau \rangle$ for every irreducible representation σ of D_{2m}.

(2) Consider D_{2m} for m even. Then (cf. Example 3.4.1) we have

$$C_1 = \{1\}, \quad C_2 = \{x, x^{m-1}\}, \dots, \quad C_{m/2} = \{x^{\frac{m}{2}-1}, x^{\frac{m}{2}+1}\},$$
$$C_{\frac{m}{2}+1} = \{x^{\frac{m}{2}}\}, \quad C_{\frac{m}{2}+2} = \{x^i y : i \text{ is even}\}, \quad C_{\frac{m}{2}+3} = \{x^i y : i \text{ is odd}\}.$$

Again, $\gamma_1 = \gamma_{\frac{m}{2}+1} = \tau$ and $\gamma_i = \tau \oplus \psi_{+-}$ for $i = 2, \dots, m/2$ (as conjugation by x is trivial on C_i but conjugation by y is not). We will determine the last two representations by computing their characters. Of course, 1 acts trivially, as does $x^{m/2}$, being central, so

$$\chi_{\frac{m}{2}+2}(1) = \chi_{\frac{m}{2}+3}(1) = \chi_{\frac{m}{2}+2}(x^{m/2}) = \chi_{\frac{m}{2}+3}(x^{m/2}) = m/2.$$

Also, $x^j(x^i y)x^{-j} = x^{i+2j}y$, so for $j \neq m/2$, x^j fixes no element and

$$\chi_{\frac{m}{2}+2}(x^j) = \chi_{\frac{m}{2}+3}(x^j) = 0 \quad \text{for } j \neq m/2.$$

Now $y(x^i y)y^{-1} = yx^i = x^{-1}y$, so y fixes $x^i y$ when $2i = 0$, i.e., $i = 0$ or $m/2$. Hence, if $m/2$ is odd,

$$\chi_{\frac{m}{2}+2}(y) = \chi_{\frac{m}{2}+3}(y) = 1,$$

while if $m/2$ is even

$$\chi_{\frac{m}{2}+2}(y) = 2 \quad \text{and} \quad \chi_{\frac{m}{2}+3}(y) = 0.$$

Similarly, if $m/2$ is odd,

$$\chi_{\frac{m}{2}+2}(xy) = \chi_{\frac{m}{2}+3}(xy) = 1,$$

while if $m/2$ is even,

$$\chi_{\frac{m}{2}+2}(xy) = 0 \qquad \text{and} \qquad \chi_{\frac{m}{2}+3}(xy) = 2.$$

Now by computation with the irreducible characters of D_{2m} (which may be read off from Example 3.1.4 (7)) we see the following.

For $m/2$ odd,

$$\gamma_{\frac{m}{2}+2} = \gamma_{\frac{m}{2}+3} = \psi_{++} \oplus \gamma'.$$

For $m/2$ even,

$$\gamma_{\frac{m}{2}+2} = \psi_{++} \oplus \psi_{-+} \oplus \gamma'$$

and

$$\gamma_{\frac{m}{2}+3} = \psi_{++} \oplus \psi_{--} \oplus \gamma',$$

where

$$\gamma' = \varphi_2 \oplus \varphi_4 \oplus \varphi_6 \oplus \cdots \oplus \varphi_k,$$

with $k = m/2 - 1$ for $m/2$ odd and $k = m/2 - 2$ for $m/2$ even.

(3) Consider A_4. We adopt the notation of Example 3.4.3. The conjugacy classes of A_4 are $C_1 = \{1\}$, $C_2 = \{I, J, K\}$, $C_3 = \{T, TI, TJ, TK\}$, and $C_4 = \{T^2, T^2I, T^2J, T^2K\}$. The centralizers of representatives of conjugacy classes (other than $\{1\}$) are $C(I) = V$, $C(T) = C(T^2) = S$, and we have computed $\mathrm{Ind}_V^G(\tau)$ and $\mathrm{Ind}_S^G(\tau)$ in Example 1.18, so we see

$$\gamma_1 = \tau,$$
$$\gamma_2 = \tau \oplus \pi^*(\theta_1) \oplus \pi^*(\theta_2),$$
$$\gamma_3 = \gamma_4 = \tau \oplus \alpha.$$

Theorem 3.24 (Brauer). *Let $\widetilde{G}$ be a group which acts on $\{C_i\}$, the set of conjugacy classes of elements of G, and also on $\{\chi_i\}$, the set of characters of irreducible complex representations of G. Denote these permutation representations by γ and γ', respectively, and suppose they satisfy*

$$\left(\gamma'(\widetilde{g})(\chi_i)\right)\left(\gamma(\widetilde{g})(c_j)\right) = \chi_i(c_j)$$

for every $\widetilde{g} \in \widetilde{G}$, and every i and j (where $c_j \in C_j$). Then γ and γ' are isomorphic as complex representations of $\widetilde{G}$.

Proof. Let $A = [a_{ij}] = [\chi_i(c_j)]$ be the character table of G, regarded as a matrix, and recall from Lemma 3.5.12 that A is nonsingular.

Let $M(\widetilde{g})$ be the matrix of $\gamma(\widetilde{g})$ in the basis $\{C_j\}$, so $M(\widetilde{g}) = [m_{ij}(\widetilde{g})]$, where $m_{ij}(\widetilde{g}) = 1$ if $\gamma(\widetilde{g})(C_i) = C_j$ and 0 otherwise.

Let $M'(\widetilde{g})$ be the matrix of $\gamma'(\widetilde{g})$ in the basis $\{\sigma_i\}$, so $M'(\widetilde{g}) = [m'_{ij}(\widetilde{g})]$, where $m'_{ij}(\widetilde{g}) = 1$ if $\gamma'(\widetilde{g})(\chi_i) = \chi_j$ and 0 otherwise.

Now the (i,j) entry of $AM(\widetilde{g})$ is

$$\sum_k \chi_i(c_k)m_{kj}(\widetilde{g}) = \chi_i(\gamma(\widetilde{g})^{-1}(c_j))$$

as for fixed j, $m_{kj}(\widetilde{g}) = 1$ when $\gamma(\widetilde{g})(C_k) = C_j$ and $= 0$ otherwise.

On the other hand, the (i,j) entry of $M'(\widetilde{g})A$ is

$$\sum_k m'_{ik}(\widetilde{g})\chi_k(c_j) = \gamma'(\widetilde{g})(\chi_i)(c_j))$$

as for fixed i, $m'_{ik}(\widetilde{g}) = 1$ when $\gamma'(\widetilde{g})(\chi_i) = \chi_k$ and $= 0$ otherwise.

But by hypothesis,

$$\chi_i(\gamma(\widetilde{g})^{-1}(c_j)) = (\gamma'(\widetilde{g})(\chi_i))(c_j)$$

for every $\widetilde{g}$, i, and j.

In other words $AM(\widetilde{g}) = M'(\widetilde{g})A$, i.e., $M(\widetilde{g}) = A^{-1}M'(\widetilde{g})A$, for every $\widetilde{g} \in \widetilde{G}$, so γ and γ' are isomorphic as complex representations of $\widetilde{G}$. $\square$

Corollary 3.25.

(1) *For each $\widetilde{g} \in \widetilde{G}$, the number of conjugacy classes fixed by $\gamma(\widetilde{g})$ and the number of characters of irreducible representations fixed by $\gamma'(\widetilde{g})$ are equal.*

(2) *The number of orbits of $\widetilde{G}$ on $\{C_j\}$ and the number of orbits of $\widetilde{G}$ on $\{\chi_i\}$ are equal.*

Proof. (1) is just the equality $\chi_\gamma(\widetilde{g}) = \chi_{\gamma'}(\widetilde{g})$ and (2) follows from (1) by the Cauchy-Frobenius theorem, Theorem 3.9. $\square$

Remark.

(1) Let $\widetilde{G}$ act on G by Γ, and suppose this action preserves conjugacy. Let $\widetilde{G}$ act on the set of irreducible representations of G by Γ', and suppose this action preserves isomorphism. Suppose further that

$$(\Gamma'(\widetilde{g})(\sigma))(\Gamma(\widetilde{g})(g)) = \sigma(g)$$

for every $\widetilde{g} \in \widetilde{G}$, every $g \in G$, and every irreducible representation σ of G. Then Γ and Γ' yield actions γ and γ' satisfying the hypotheses of Theorem 3.24.

(2) As a special case of Theorem 3.24, we have the situation where G is a normal subgroup of $\widetilde{G}$, Γ is the action of $\widetilde{G}$ on G by conjugation, and Γ' is the action $\sigma \mapsto \sigma^{\tilde{g}}$. (Recall that $\sigma^{\tilde{g}}$ denotes the conjugate of σ by $\tilde{g}$.) Note in this case the action of $\widetilde{G}$ factors through $\widetilde{G}/G$.

(3) Let F be any group of automorphisms of G. Let Γ be the action of F on G, i.e., $\Gamma(f)(g) = f(g)$, for $f \in F$, and define Γ' by $\Gamma'(f)(\sigma)(g) = \sigma(\Gamma(f)^{-1}(g))$. Then Theorem 3.24 applies to this situation. Note, however, that in this situation we may form the semidirect product

$$1 \to G \to \widetilde{G} \to F \to 1,$$

and we are in the situation of (2) (as in $\widetilde{G}$, $(1, f)(g, 1)(1, f)^{-1} = (f(g), 1)$).

(4) Note that, in (3), if F' is the subgroup of F consisting of those f that are inner automorphisms of G, then the action of F factors through F/F'. Thus we do not lose any information if instead we form the semidirect product

$$1 \to G \to \widetilde{G} \to F/F' \to 1.$$

Example 3.26. Note that $\mathbb{Z}/m\mathbb{Z}$ acts as a group of automorphisms of D_{2m}, where a generator f of $\mathbb{Z}/m\mathbb{Z}$ acts by $f(s) = x$, $f(y) = xy$. When m is odd, f is just conjugation by $x^{(m+1)/2}$, so γ and γ' (as defined in the previous remark) are both trivial. However, when m is even, f is not an inner automorphism. Note that in this case γ leaves all conjugacy classes fixed except that it interchanges $C_{\frac{m}{2}+2}$ and $C_{\frac{m}{2}+3}$, and γ' leaves all isomorphism classes of irreducible representations fixed except that it interchanges ψ_{-+} and ψ_{--}. (Compare the results of Example 3.23 (2).)

4.4. M-groups

In this section all representations are over the complex numbers.

Recall that a monomial representation of a group G is a representation isomorphic to $\mathrm{Ind}_H^G(\sigma)$ for some (not necessarily proper) subgroup H of G and some 1-dimensional representation σ of H.

Definition 4.1. G is an M-**group** if every irreducible complex representation of G is a monomial representation.

We have already seen a number of examples of M-groups. Every abelian group is, of course, an M-group. Every dihedral group D_{2m} is an M-group, as are the quaternion group Q_8 and the alternating group A_4. Mackey's construction (Construction 2.13 and Theorem 2.14) shows that every finite group $G = AH$ that is the semidirect product of a normal abelian subgroup A and an abelian subgroup H is an M-group.

Our objective in this section, both for its own interest and for its use in the next section, will be to develop criteria that ensure that a given group is an M-group.

We have the following general result.

Theorem 4.2.

(1) *Let $G = G_1 \times G_2$. Then every irreducible representation of G is isomorphic to $V_1 \otimes_{\mathbf{C}} V_2$ for some irreducible representations V_i of G_i, $i = 1, 2$, and every representation of G of this form is irreducible.*

(2) *If $V_i = \mathrm{Ind}_{H_i}^{G_i}(W_i)$, $i = 1, 2$, then $V_1 \otimes_{\mathbf{C}} V_2 = \mathrm{Ind}_{H_1 \times H_2}^{G_1 \times G_2}(W_1 \otimes_{\mathbf{C}} W_2)$.*

(3) *If $K \subseteq H \subseteq G$ with K a normal subgroup of G, and W is a representation of H/K, then $\pi^*(\mathrm{Ind}_{H/K}^{G/K}(W)) = \mathrm{Ind}_H^G(\pi^*(W))$, where π denotes both quotient maps $G \to G/K$ and $H \to H/K$.*

Proof. Straightforward; we leave it to the reader. $\qquad\qquad\square$

Corollary 4.3. *If G_1 and G_2 are M-groups, then $G_1 \times G_2$ is also an M-group.*

Proof. $\qquad\qquad\square$

We now introduce several classes of groups.

Definition 4.4.

(1) A group G is **solvable** if it has a composition series, i.e., if there is a sequence of subgroups $\{1\} = G_0 \subset G_1 \subset \cdots \subset G_k = G$ with G_{i-1} a normal subgroup of G_i and with quotient G_i/G_{i-1} abelian, for each $i = 1, \ldots, k$.

(2) A group G is **supersolvable** if G_{i-1} is normal in G and G_i/G_{i-1} is cyclic, for each $i = 1, \ldots, k$.

(3) A group G is **nilpotent** if G_{i-1} is normal in G and G_i/G_{i-1} is a subgroup of the center of G/G_{i-1}, for each $i = 1, \ldots, k$.

Note that nilpotent implies supersolvable (as if G_i/G_{i-1} is not cyclic, choose $G_{i-1} \subset G_i' \subset G_i$ with G_i'/G_{i-1} cyclic, and it is easy to check that G_i' is normal in G) and supersolvable trivially implies solvable.

Lemma 4.5. *Every p-group is nilpotent.*

Proof. Let G be a p-group. Recall that the center of G is nontrivial. (Proof: Write G as a disjoint union of conjugacy classes. The order of G is a positive power of p and the order of every conjugacy class is either 1 or a positive power of p, so the number of conjugacy classes consisting of one element is a multiple of p. But this number is the order of the center of G.)

We argue by induction on the order of G.

Let C be the center of G. If $C = G$, then G is abelian and we are done, so suppose not. Let $\pi : G \to G/C$. Now G/C is a p-group, and so is nilpotent; let $\{1\} \subset \overline{H}_1 \subset \cdots \subset \overline{H}_k = G/C$ be a composition series for G/C, and let $H_i = \pi^{-1}(\overline{H}_i)$. Then $\{1\} \subset C \subset H_1 \subset \cdots \subset H_k = G$ is a composition series for G, and G is nilpotent. $\qquad\square$

Definition 4.6.

(1) A group G is **metacyclic** if it has a cyclic normal subgroup G_1 with G/G_1 cyclic.

(2) A group G is **metabelian** if it has an abelian normal subgroup G_1 with G/G_1 abelian.

Note that metacyclic implies nilpotent and metabelian implies solvable.

We use the following technical result.

Lemma 4.7. *Let G be a finite group and suppose that G has a normal abelian subgroup A not contained in the center of G. Then every faithful irreducible representation V of G is of the form $\mathrm{Ind}_H^G(W)$ for some proper subgroup H of G and some irreducible representation W of H.*

Proof. Let V be given by $\sigma : G \to \mathrm{Aut}(V)$. By Burnside's theorem, Theorem 3.3.8, $\{\sigma(g)\}$ span $\mathrm{End}_{\mathbf{C}}(V)$, whose center consists of homotheties. Since σ is faithful and A is not contained in the center of G, there is some $a \in A$ with $\sigma(a)$ not a homothety. Hence $\mathrm{Res}_A^G(V)$ is not isotypic, so, by Corollary 1.13, $V = \mathrm{Ind}_H^G(W)$ for some proper subgroup H of G (containing A) and some irreducible representation W of H. $\qquad\square$

Theorem 4.8.

(1) *Let G be a supersolvable group. Then G is an M-group.*

(2) *Let G be a metabelian group. Then G is an M-group.*

Proof. If G is abelian, the theorem is trivial, so suppose not. First observe that any subgroup or quotient group of a supersolvable (resp. metabelian) group is supersolvable (resp. metabelian).

Let G be supersolvable or metabelian.

We argue by induction on the order of G. Let $\sigma : G \to \mathrm{Aut}(V)$ be a representation of G. First suppose σ is not faithful. Let $K = \mathrm{Ker}(\sigma)$, $\overline{G} = G/K$, $\pi : G \to \overline{G}$ be the quotient map, and $\overline{\sigma} : \overline{G} \to \mathrm{Aut}(V)$ with $\sigma = \overline{\sigma}\pi = \pi^*(\overline{\sigma})$. Then $\overline{G}$ is supersolvable or metabelian, so by the inductive hypothesis $\overline{\sigma} = \mathrm{Ind}_{\overline{H}}^{\overline{G}}(\overline{\rho})$ for some subgroup $\overline{H}$ of $\overline{G}$ and some 1-dimensional representation $\overline{\rho}$ of $\overline{H}$. Let $H = \pi^{-1}(\overline{H})$ and let $\rho = \overline{\rho}\pi = \pi^*(\overline{\rho})$. Then $\sigma = \mathrm{Ind}_H^G(\rho)$ by Theorem 4.2 (3).

Thus we are reduced to the case when σ is faithful. We claim that in this case G satisfies the hypothesis of Lemma 4.7. Thus $\sigma = \mathrm{Ind}_H^G(\rho)$ for some subgroup H and some irreducible representation ρ of H. Then H is supersolvable or metabelian, so $\rho = \mathrm{Ind}_K^H(\varphi)$ for some 1-dimensional representation φ of K. Then $\sigma = \mathrm{Ind}_K^G(\varphi)$ as required.

It remains to prove the claim, in each of the cases of the theorem.

(1) Let C be the center of G. If $C = \{1\}$, then let $A = G_1$, and we are done. If $C \neq \{1\}$, consider $H = G/C$; let $\pi = G \to H$. H is supersolvable; let H_1 be the first group in its composition series, and let $A = \pi^{-1}(H_1)$. We claim A is as required. Since H_1 is normal in H, A is certainly normal in G. We must show A is abelian. Let $\overline{h}$ be a generator of H_1 and let $h \in G$ with $\pi(h) = \overline{h}$. Then any element in A can be written in the form $h^i c$ for some i and for some $c \in C$. Then $(h^{i_1}c_1)(h^{i_2}c_2) = h^{i_1}h^{i_2}c_1 c_2 = h^{i_2}h^{i_1}c_2 c_1 = (h^{i_2}c_2)(h^{i_1}c_1)$, so A is abelian.

(2) Let A' be an abelian normal subgroup of G, with G/A' abelian, and let C be the center of G. If A' is not contained in C, we are done, so suppose A' is contained in C. Then G/C is a quotient of G/A', which is abelian. Let H_1 be a cyclic subgroup of G/C and let $A = \pi^{-1}(H_1)$, where $\pi : G \to G/C$. Then A is an abelian normal subgroup of G as in part (1). $\square$

In the next section, we will use the following special case of this theorem.

Corollary 4.9. *Every p-group is an M-group.*

Proof. This follows immediately from Lemma 4.5 and Theorem 4.8. $\square$

4.5. Theorems of Artin and Brauer

In this section we first prove Artin's theorem on expressing characters of G in terms of characters induced from cyclic subgroups of G, and apply it to complex representations of G whose characters are $\mathbf{Q}$-valued. We then prove Brauer's theorem on expressing characters of G in terms of characters induced from "elementary" subgroups of G, following the argument of Brauer and Tate [BT], and use it, as they did, to prove that every complex representation of G is defined over $\mathbf{Q}(\sqrt[n]{1})$.

Lemma 5.1. *Let $\mathbf{F}$ be a subfield of $\mathbf{C}$. Let G be a cyclic group and let $\sigma : G \to \mathrm{Aut}(V)$ be a complex representation whose character $\chi = \chi_\sigma$ is $\mathbf{F}$-valued. Then σ is defined over $\mathbf{F}$.*

Proof. Pick a generator g of G. Then $\chi(g) = \zeta_1 + \cdots + \zeta_d$, where ζ_i is an n-th root of unity, $i = 1, \ldots, d = \deg(\sigma)$, and V has a basis in which $[\sigma(g)] = D = \operatorname{diag}(\zeta_1 \ldots, \zeta_d)$. The characteristic polynomial of D is $f(x) = (x - \zeta_1) \cdots (x - \zeta_d)$. But D is similar to $C(f(x))$, the companion matrix of $f(x)$. The fact that $\chi(g^m) \in \mathbf{F}$ for every positive integer m readily implies (as the coefficients of $C(f(x))$ are, up to sign, the elementary symmetric polynomials in $\zeta_1, \ldots, \zeta_d$) that all of the coefficients of $f(x)$ are in $\mathbf{F}$, so all of the entries of $C(f(x))$ are elements of $\mathbf{F}$, and so σ is defined over $\mathbf{F}$. $\square$

Lemma 5.2. *Let $\sim$ be the equivalence relation on G defined by $g_1 \sim g_2$ if g_1 and g_2 generate conjugate (cyclic) subgroups of G. Let $\{H_i\}$, $i \in I$, be a maximal set of pairwise nonconjugate cyclic subgroups of G. Let*

$$\varphi_i = \frac{1}{[N(H_i) : H_i]} \operatorname{Ind}_{H_i}^G (1).$$

Then $\{\varphi_i\}$ is a basis of the $\mathbb{Z}$-lattice

$$T = \{f : G \to \mathbb{Z} \mid f \text{ is constant on } \sim \text{ equivalence classes}\}.$$

Proof. First observe, from Corollary 1.22, that:

(1) If $g \in G$ is not conjugate to an element of H_i, then $\varphi_i(g) = 0$.

(2) If $g \in G$ is conjugate to a generator of H_i, then $\varphi_i(g) = 1$.

(3) $\varphi_i(g) \in \mathbb{Z}$ for every $g \in G$.

Next observe that $g_1 \sim g_2$ if and only if g_1 is conjugate to g_2^m for some m relatively prime to the order of g_1. Then, since φ_1 is a rational multiple of the character of a rational representation of G, Theorem 3.7.1 (2) implies

(4) φ_i is constant on $\sim$ equivalence classes.

In particular, by (3) and (4), $\varphi_i \in T$ for each i.

For each $i \in I$, let

$$K_i = \{g \in G \mid \text{the subgroup generated by } g \text{ is conjugate to } H_i\}.$$

Then $\{K_i\}$ are the $\sim$ equivalence classes. Define $\theta_i : G \to \mathbb{Z}$ by

$$\theta_i(g) = \begin{cases} 1 & \text{if } g \in K_i, \\ 0 & \text{if } g \notin K_i. \end{cases}$$

Then $\{\theta_i\}$, $i \in I$, is clearly a basis for T. Thus to prove the lemma it suffices to prove that $\{\theta_i\}$ are expressible as $\mathbb{Z}$-linear combinations of $\{\varphi_i\}$.

Order I so that H_j conjugate to a subgroup of H_i implies $j \leq i$.

If $g \in H_i$, then g generates a subgroup of H_i, so $g \in K_j$ for some $j \leq i$, and $g \in K_i$ if and only if g is a generator of H_i.

Since $\{\theta_i\}$ is a basis, we have

$$\varphi_i = \sum_{i \in I} p_{ij}\theta_j \quad \text{for each } i \in I,$$

for some integers p_{ij}. If $g \in K_i$, then g is conjugate to a generator of H_i, so

$$1 = \varphi_i(g) = p_{ii}\theta_i(g) = p_{ii}.$$

If $g \in K_j$ for some $j > i$, then the subgroup generated by g is not conjugate to a subgroup of H_i, so g is not conjugate to an element of H_i, and

$$0 = \varphi_i(g) = p_{ij}\theta_j(g) = p_{ij}.$$

Thus the matrix $P = (p_{ij})$ that expresses $\{\varphi_i\}$ in terms of $\{\theta_i\}$ is a triangular matrix with integer entries and with all diagonal entries equal to 1. Then P^{-1} is also an integer matrix (triangular with diagonal entries equal to 1). But P^{-1} expresses $\{\theta_i\}$ in terms of $\{\varphi_i\}$, so each θ_i is in the integer span of $\{\varphi_i\}$, and we are done. $\qquad\square$

Theorem 5.3 (Artin).

(1) *Let σ be a complex representation of G such that $\chi(g) = \chi_\sigma(g)$ is $\mathbf{Q}$-valued for every $g \in G$. Let $\{H\}$ be the set of cyclic subgroups of G, and let ψ_H be the character of the representation $\mathrm{Ind}_H^G(\tau)$. Then*

$$\chi = \sum_H \frac{a_H}{[N(H) : H]}\psi_H$$

for some integers $\{a_H\}$. In particular,

$$n\chi = \sum_H b_H\psi_H$$

for some integers $\{b_H\}$.

(2) *Let $\mathbf{F}$ be a subfield of $\mathbf{C}$, and let σ be a complex representation of G such that $\chi(g) = \chi_\sigma(g)$ is $\mathbf{F}$-valued for every $g \in G$. Let $\{H\}$ be the set of cyclic subgroups of G. Then for each H there is an $\mathbf{F}$-representation ρ_H of H such that, if ψ'_H is the character of $\mathrm{Ind}_H^G(\rho_H)$, then*

$$n\chi = \sum_H b_h\psi'_H$$

for some integers $\{b_H\}$.

Proof.

(1) For each $g \in G$, $\chi_\sigma(g)$ is an element of $\mathbf{Q}$, and is an algebraic integer, so is an element of $\mathbb{Z}$. Also, χ_σ is constant on $\sim$ equivalence classes, so this theorem immediately follows from Lemma 5.2.

(2) By part (1),

$$n\chi_\tau = \sum_H b_H \psi_H$$

for some integers $\{b_H\}$. Of course, $n\chi_\tau$ is just the constant function $n\chi_\tau(g) = n$ for every $g \in G$. Then

$$
\begin{aligned}
n\chi &= \left(\sum_H b_H \psi_H\right)\chi \\
&= \left(\sum_H b_H \operatorname{Ind}_H^G(1)\right)\chi \\
&= \sum_H b_H \left(\operatorname{Ind}_H^G(1)\chi\right) \\
&= \sum_H b_H \operatorname{Ind}_H^G(\operatorname{Res}_H^G(\chi))
\end{aligned}
$$

by Lemma 1.23.

Now $\operatorname{Res}_H^G(\chi)$ is an $\mathbf{F}$-valued character of H, so by Lemma 5.1 it is the character of an $\mathbf{F}$-representation ρ_H of H, completing the proof. $\qquad\square$

Remark. Note that the $\{a_H\}$ (or $\{b_H\}$) are not necessarily positive. An example is provided by the representation $\mathcal{R}_0$ (see Example 3.1.7) of $G = \mathbb{Z}/p\mathbb{Z}$. Recall that $\mathcal{R} \cong \mathcal{R}_0 \oplus \tau$, so if χ is the character of $\mathcal{R}_0$, then, in the notation of Theorem 5.3 (1), $\chi = \psi_{\{1\}} - \psi_G$.

Lemma 5.2 also allows us to generalize Lemma 3.7.

Corollary 5.4. *Let $\sigma_i : G \to \operatorname{Aut}(V_i)$, $i = 1, 2$, be two complex representations of G such that χ_{σ_1} and χ_{σ_2} are $\mathbf{Q}$-valued. Then V_1 and V_2 are equivalent if and only if*

$$\dim_{\mathbf{C}} \{v \in V_1 \mid \sigma_1(g)(v) = v\} = \dim_{\mathbf{C}} \{v \in V_2 \mid \sigma_2(g)(v) = v\}$$

for every $g \in G$.

Proof. The only-if part is trivial. For the if part, we follow the notation of the proof of Lemma 5.2. Note that the condition in the statement of the corollary is equivalent to

$$\langle \tau, \operatorname{Res}_H^G(V_1)\rangle = \langle \tau, \operatorname{Res}_H^G(V_2)\rangle$$

for every cyclic subgroup H of V. By Frobenius reciprocity, this is equivalent to

$$\langle V_1, \mathrm{Ind}_H^G(\tau) \rangle \;=\; \langle V_2, \mathrm{Ind}_H^G(\tau) \rangle$$

for every cyclic subgroup H. Expressing this in terms of characters yields

$$\langle \chi_{\sigma_1}, \varphi_i \rangle \;=\; \langle \chi_{\sigma_2}, \varphi_i \rangle$$

for each $i \in I$, which then yields

$$\langle \chi_{\sigma_1}, \theta_i \rangle \;=\; \langle \chi_{\sigma_2}, \theta_i \rangle$$

for each $i \in I$, so $\chi_{\sigma_1} = \chi_{\sigma_2}$ and V_1 and V_2 are equivalent. $\square$

Now we turn to Brauer's theorem.

Lemma 5.5. *Fix a prime number p. Then any $g \in G$ can be written uniquely as $g = ab$, where the order of a is prime to p, the order of b is a power of p, and a and b commute. In this situation both a and b are powers of g.*

Proof. If $(n, p) = 1$, the proof is trivial. If $n = p^r q$ with $r \geq 1$ and $(p, q) = 1$, then set $1 = xp^r + yq$ for some integers x and y. Set $a = g^{xp^r}$ and $b = g^{yq}$. Uniqueness is easy to check. $\square$

Definition 5.6. In the situation of Lemma 5.5, a is the **p-regular factor** of g and b is the **p-singular factor** of g. If $b = 1$, i.e., if the order of g is prime to p, then g is a **p-regular element** of G. If $a = 1$, i.e., if the order of g is a power of p, then g is a **p-singular element** of G.

Definition 5.7.

(1) Fix a prime number p. A subgroup of G is **p-elementary** if it is of the form $A \times B$, where A is a cyclic group of order prime to p and B is a p-group.

(2) A subgroup of G is **elementary** if it is p-elementary for some prime p.

Definition 5.8.

(1) Let $\sigma_1, \ldots, \sigma_t$ be the irreducible complex representations of G. The (complex) **representation ring** $K(G)$ is the free abelian group with generators $\sigma_1, \ldots, \sigma_t$, with multiplication defined by the tensor product of representations. An element of $K(G)$ is called a **virtual representation**.

(2) Let $\chi_1, \ldots, \chi_t$ be the characters of the irreducible complex representations of G. The (complex) **character ring** $X(G)$ is the free abelian group with generators $\chi_1, \ldots, \chi_t$, with multiplication defined by the product of characters (viewed as functions on G). An element of $X(G)$ is called a **virtual character**.

From the fact that $(\chi_{\sigma_1})(\chi_{\sigma_2}) = \chi_{\sigma_1 \otimes \sigma_2}$, we see that $X(G)$ is closed under multiplication, and indeed that the map $\sigma \mapsto \chi_\sigma$ gives an isomorphism between $K(G)$ and $X(G)$. (The reader familiar with K-theory will recognize that $K(G)$ is $K(\mathbf{C}(G))$.)

We now need to make some technical definitions.

Definition 5.9.

(1) Let I be a subring of $\mathbf{C}$ that is of finite rank over $\mathbb{Z}$. Then $X_I = I \otimes_\mathbb{Z} X(G)$.

(2) Let $\{H_\delta\}_{\delta \in \Delta}$ be a fixed family of subgroups of G. For each $\delta \in \Delta$, let $\{\rho_{\delta_j}\}$ be the set of irreducible complex representations of H_δ, and let $\{\psi_{\delta_j}\}$ be the corresponding characters. Then $V(G)$ is the $\mathbb{Z}$-span of the characters $\{\operatorname{Ind}_{H_\delta}^G(\psi_{\delta_j})\}$, and $V_I = I \otimes_\mathbb{Z} V(G)$.

(3) Let U_I be the subring of the ring of class functions on G given by

$$U_I = \{\text{class functions } \theta : G \to \mathbf{C} \mid \operatorname{Res}_{H_\delta}^G(\theta) \in X_I(H_\delta) \text{ for each } \delta \in \Delta\}.$$

(4) Let U, V, and X denote U_I, V_I, and X_I for $I = \mathbb{Z}$.

Our goal is to prove Theorems 5.19 and 5.21. Because of the length of the proof, we pause here to describe its strategy. The key to the proof is to show that if $\{H_\delta\}$ is the set of elementary subgroups of V, then $1 \in V$ (see the proof of Corollary 5.18). Along the way we show in Corollary 5.15 that $n \in V$ in this case. Actually we already know this, as we have shown in Theorem 5.3 that it is true when $\{H_\delta\}$ is the set of cyclic subgroups (and every cyclic subgroup is elementary). However, we follow Brauer and Tate's proof as Corollary 5.15 is a more general result, and we need Lemma 5.13 in any case. Also, while it is V we are interested in, V_I is easier to work with, but we show in Lemma 5.11 that any integer in V_I is actually in V.

Remarks.

(1) Through Corollary 5.18, we let I be as in Definition 5.9 (1), though we will soon specialize I further.

(2) We embed I in X_I by $1 \mapsto \tau \mapsto \chi_\tau$, and regard I as a subring of X_I.

(3) By Frobenius reciprocity, $\theta \in U_I$ if and only if $\langle \theta, \operatorname{Ind}_{H_\delta}^G(\psi_{\delta_j}) \rangle \in I$ for every ψ_{δ_j}.

Proposition 5.10. *U_I is a ring, V_I is an ideal of U_I, and*

$$U_I \supseteq X_I \supseteq V_I.$$

Proof. Clearly U_I is a ring, and clearly $U_I \supseteq X_I$, as every character of G restricts to a character of H_δ for every δ. Also, clearly $X_I \supseteq V_I$ as V is spanned by the characters of certain representations of G. To show that V_I is an ideal of U_I, it suffices to show that $\theta \operatorname{Ind}_{H_\delta}^G(\psi_{\delta_j}) \in V_I$ for each $\theta \in U_I$ and each ψ_{δ_j}. But

$$\theta \operatorname{Ind}_{H_\delta}^G(\psi_{\delta_j}) = \operatorname{Ind}_{H_\delta}^G(\operatorname{Res}_{H_\delta}^G(\theta)\psi_{\delta_j}) \in V_I$$

by Lemma 1.23. $\square$

Lemma 5.11. $V_I \cap \mathbb{Z} = V \cap \mathbb{Z}$.

Proof. Since I is a $\mathbb{Z}$-module of finite rank, I has a basis $\{\eta_\mu\}$ where one of the basis elements $\eta_{\mu_0} = 1$. (Note 1 is a primitive element of I, as otherwise I would not be finitely generated over $\mathbb{Z}$. Then this claim follows from [AW, Theorem 3.6.16].)

We claim that $\{\eta_\mu\}$, as elements of X_I, are X-linearly independent. Suppose $\sum_\mu \eta_\mu \sum_i c_{\mu i} \chi_i = 0$ with $c_{\mu i} \in \mathbb{Z}$. Then $\sum_i (\sum_\mu \eta_\mu c_{\mu i} \chi_i) = 0$, so by independence of characters, $\sum_\mu \eta_\mu c_{\mu i} = 0$ for each i, and then $c_{\mu i} = 0$ for each μ and each i.

Now suppose $d \in V_I \cap \mathbb{Z}$. Then, for some $i_{\delta_j} \in I$,

$$d\eta_{\mu_0} = d \cdot 1 = d = \sum_{\delta,j} i_{\delta_j} \operatorname{Ind}_{H_\delta}^G(\psi_{\delta_j}) = \sum_{\delta,j} \left(\sum_\mu z_{\delta_{j\mu}} \eta_\mu \right) \operatorname{Ind}_{H_\delta}^G(\psi_{\delta_j})$$

with $z_{\delta_{j\mu}} \in \mathbb{Z}$, and comparing the coefficients of $\eta_{\mu_0} = 1$ on each side, we see $d \in V$. $\square$

Henceforth, we choose $I = \mathbb{Z}(\sqrt[n]{1})$, and observe that all values of all characters of all representations of all subgroups of G lie in I.

Lemma 5.12. *Fix a prime p, and write $g = ab$ as in Lemma 5.5. For any $\theta \in X_I$ with $\theta(g) \in \mathbb{Z}$ and $\theta(a) \in \mathbb{Z}$, $\theta(g) \equiv \theta(a) \pmod{p}$.*

Proof. It suffices to prove this when G is cyclic, generated by g. Then the irreducible representations of G are 1-dimensional, so the irreducible characters of G are homomorphisms: $\chi_i : G \to \mathbf{C}^*$. Thus, if b has order $q = p^r$, $\chi_i(g)^q = \chi_i(a)^q$. Now $\theta = \sum \alpha_i \chi_i$ with $\alpha_i \in I$. Raising this equation to the p-th power r times in succession, and using the multinomial theorem, we obtain

$$\theta(g)^q \equiv \sum_i \alpha_i^q \chi_i(g)^q = \sum_i \alpha_i^q \chi_i(a)^q \equiv \theta(a)^q \pmod{pI}.$$

Since $pI \cap \mathbb{Z} = p\mathbb{Z}$, this gives $\theta(g)^q \equiv \theta(a)^q \pmod{p}$. But now, by Fermat's theorem, $z^q \equiv z \pmod{p}$ for any integer z, proving the lemma. $\square$

Lemma 5.13. *Suppose that some group H_δ in $\{H_\delta\}$ contains a subgroup $H = A \times B$, the direct product of an abelian group A and a group B, with $(|A|, |B|) = 1$. Fix $a \in A$ and let $C(a)$ be the centralizer of a in G, $C(a) = \{g \in G \mid ga = ag\}$. Then there exists an element $\varphi \in V_I$ satisfying:*

(1) $\varphi(g) \in \mathbb{Z}$ for every $g \in G$.

(2) $\varphi(g) = 0$ if g is not conjugate to an element of aB.

(3) $\varphi(a) = [C(a) : B]$.

Proof. We construct an element $\psi \in X_I(H)$ such that $\varphi = \mathrm{Ind}_{H_\delta}^G(\mathrm{Ind}_H^{H_\delta} \psi)$ has the desired properties.

Composing an irreducible (i.e., 1-dimensional) representation of A with the projection $A \times B \to A$ gives a representation of H; let $\{\omega_i\}$ be the set of $|A|$ characters of representations of H obtained in this way. The desired element ψ is given by

$$\psi = \sum_i \overline{\omega}_i(a)\omega_i.$$

(Note that $\overline{\omega}_i(a) \in I$ so $\psi \in X_I(H)$.)

The orthogonality relations for characters show that $\psi(h) = |A|$ for $h \in aB$ and $\psi(h) = 0$ for $h \in H$, $h \notin aB$. Then Theorem 1.20 shows that $\varphi(g) \in \mathbb{Z}$ for all $g \in G$, and that $\varphi(g) = 0$ if g is not conjugate to an element of aB.

We also use Theorem 1.20 to compute $\varphi(a)$. We have, in the notation of Theorem 1.20, that $\widetilde{\psi}(qaq^{-1}) = |A|$ if $qaq^{-1} \in aB$ and $\widetilde{\psi}(qaq^{-1}) = 0$ if not. Now $qaq^{-1} \in aB$ implies $qaq^{-1} = a$, as otherwise the order of qaq^{-1} would not divide $|A|$, so $qaq^{-1} \in aB$ if and only if $q \in C(a)$, and then Corollary 1.21 yields $\varphi(a) = (1/|H|)|C(a)||A| = |C(a)|/|B| = [C(a) : B]$. $\qquad\square$

Lemma 5.14. *If $\bigcup_{\delta \in \Delta} H_\delta = G$, then V_I contains every class function θ on G such that $\theta(g) \in nI$ for every $g \in G$.*

Proof. Let $a \in G$ be arbitrary. Let A be the cyclic group generated by a, and let $B = \{1\}$. Then $A \times B$ is contained in some H_δ, so we may apply Lemma 5.13 to obtain a class function φ with $\varphi(g) = 0$ if g is not conjugate to a, and $\varphi(g) = |C(a)|$ if g is conjugate to a. Now $|C(a)|$ divides n, and so we may write any class function satisfying the hypothesis of the lemma as a sum of integer multiples of the functions φ obtained by letting a range over representatives of the conjugacy classes of G. $\qquad\square$

Corollary 5.15. *If $\bigcup_{\delta \in \Delta} H_\delta = G$, then $n \in V$.*

Proof. By Lemma 5.14, n (regarded as a constant function on G) lies in V_I, and $n \in \mathbb{Z}$, so $n \in V_I \cap \mathbb{Z} = V \cap \mathbb{Z}$ by Lemma 5.11. $\qquad\square$

Lemma 5.16. *If every p-elementary subgroup H of G is contained in some H_δ, then there is an element η of V_I such that $\eta(g) \in \mathbb{Z}$ for every $g \in G$ and $\eta(g) \equiv 1 \pmod{p}$ for every $g \in G$.*

Proof. Let $\{a\}$ be a set of representatives for the conjugacy classes of p-regular elements of G. For each $a \in A$ we construct $\eta_a \in V_I$ satisfying:

(1) $\eta_a(g) \in \mathbb{Z}$ for every $g \in G$.

(2) $\eta_a(g) = 0$ if the p-regular factor of g is not conjugate to a.

(3) $\eta_a(g) \equiv 1 \pmod{p}$ if the p-regular factor of g is conjugate to a.

Then $\eta = \sum \eta_a$ is the required element of V_I, as for every $g \in G$, the p-regular factor of g is conjugate to a unique a.

Fix a. We construct η_a as follows. Let A be the cyclic subgroup generated by a and let B be a p-Sylow subgroup of $C(a)$, and let $H = AB = A \times B$. Then H is p-elementary so is contained in some H_δ. Apply Lemma 5.13 to obtain an element $\varphi = \varphi_a$ of V_I satisfying the conclusion of that lemma. Since B is the p-Sylow subgroup of $C(a)$, $[C(a) : B]$ is relatively prime to p, so there is an integer z_a with $z_a[C(a) : B] \equiv 1 \pmod{p}$. Set $\eta_a = z_a\varphi_a$. Clearly η_a satisfies (1). Let $g \in G$ have the p-regular factor a'. If a' is not conjugate to a, then g is not conjugate to an element of aB and so η_a satisfies (2). Finally, if a' is conjugate to a, then $\varphi(a') = \varphi(a)$ and $1 \equiv \eta_a(a) \equiv \eta_a(g) \pmod{p}$ by Lemma 5.12, so η_a satisfies (3). $\qquad\square$

Lemma 5.17. *Fix a prime p, let $n = p^r q$ with $(p, q) = 1$, and suppose that every p-elementary subgroup of G is contained in some H_δ. Then $q \in V$.*

Proof. Let η be as in Lemma 5.16. If z is any integer with $z \equiv 1 \pmod{p^i}$, then $z^p \equiv 1 \pmod{p^{i+1}}$. Thus we have

$$\eta^{p^r}(g) \equiv 1 \pmod{p^r}$$

for every $g \in G$. Since $\eta \in V_I$ and V_I is an ideal, η^{p^r} is also an element of V_I. Let $\theta_0 = \eta^{p^r} - 1$. Then $\theta = q\theta_0$ is a class function with $\theta(g) \in n\mathbb{Z}$ for every $g \in G$, so by Lemma 5.14, $\theta(g) \in V_I$ and hence $q = q\eta^{p^r} - \theta \in V_I$ (again as V_I is an ideal). But $q \in \mathbb{Z}$, so $q \in V_I \cap \mathbb{Z} = V \cap \mathbb{Z}$ by Lemma 5.11. $\qquad\square$

Corollary 5.18. *Suppose that every elementary subgroup of G is contained in some H_δ. Then*

(1) $U = X = V$,

(2) $U_I = X_I = V_I$.

Proof.

(1) Apply Lemma 5.17 for every prime p to conclude that V contains an integer prime to p for every prime p, which implies that $1 \in V$. Since V is an ideal of U, this gives $U = X = V$.

(2) $V_I \supseteq V$, so $1 \in V_I$. Again, V_I is an ideal of U_I, so $U_I = X_I = V_I$. $\qquad\square$

Theorem 5.19 (Brauer).

(1) *Every character χ of a finite group G can be expressed as*

$$\chi = \sum a_i \operatorname{Ind}_{H_i}^{G}(\psi_i),$$

with $a_i \in \mathbb{Z}$, where ψ_i is an irreducible character of an elementary subgroup H_i of G.

(2) *A complex-valued class function θ on G belongs to the character ring $X(G)$ if and only if for every elementary subgroup H of G, $\operatorname{Res}_{H}^{G}(\theta)$ belongs to the character ring $X(H)$.*

Proof. Part (1) is just the equality $X = V$ and part (2) is just the equality $U = X$. $\square$

Lemma 5.20. *Let H be an elementary group. Then every irreducible representation of H is induced from a 1-dimensional representation of a subgroup of H.*

Proof. Let $H = A \times B$ with B a p-group and A cyclic (of order prime to p). A is an M-group, as it is abelian, and B is an M-group by Corollary 4.9, so $A \times B$ is an M-group by Corollary 4.3. $\square$

Theorem 5.21 (Brauer). *Every irreducible complex representation of G is defined over $\mathbf{Q}(\sqrt[u]{1})$, where u is the exponent of G.*

Proof. If K is any subgroup of G, then the exponent of K divides the exponent of G. If ρ is a 1-dimensional representation of K, then $\rho : K \to \mathbf{C}^*$ is a homomorphism, so $\rho(k)^u = 1$ for every $k \in K$, and ρ is defined over $\mathbf{Q}(\sqrt[u]{1})$. Then $\operatorname{Ind}_{K}^{G}(\rho)$ is also defined over $\mathbf{Q}(\sqrt[u]{1})$ in this case.

Now let σ be any irreducible representation of G. Then Theorem 5.19 (1), Lemma 5.20, and the above observation show the following.

Conclusion. χ_σ is a sum of integer multiples of the characters of representations of G defined over $\mathbf{Q}(\sqrt[u]{1})$.

Using this fact we may complete the proof in two different ways.

(1) Letting $\rho_1, \ldots, \rho_s$ be the irreducible $\mathbf{Q}(\sqrt[u]{1})$-representations of G, we have that $\chi_\sigma = \sum m_i \chi_{\rho_i}$. It remains only to show that the m_i are nonnegative, as then $\sigma = \bigoplus m_i \rho_i$.

As σ is a representation, the intertwining number $\langle \sigma, \rho_i \rangle$ is nonnegative. Also, by Lemmas 3.5.20 and 3.5.21, $\langle \rho_i, \rho_j \rangle = 0$ for $j \neq i$, and $\langle \rho_i, \rho_i \rangle$ is positive. Thus $\langle \chi_\sigma, \chi_{\rho_i} \rangle = \langle \sigma, \rho_i \rangle = m_i \langle \rho_i, \rho_i \rangle$ so m_i is nonnegative, as claimed.

(2) We have that $\chi_\sigma = \sum_{i \in I} a_i \chi_{\varphi_i}$, where φ_i are representations of G defined over $\mathbf{Q}(\sqrt[u]{1})$ and $a_i \in \mathbb{Z}$. Let $I' = \{i \mid a_i < 0\}$ and $I'' = \{i \mid a_i \geq 0\}$. Then

$$\chi_\sigma + \sum_{i \in I'} (-a_i)\chi_{\varphi_i} = \sum_{i \in I''} a_i \chi_{\varphi_i}.$$

Thus, if we let

$$\alpha = \sum_{i \in I'} (-a_i)\varphi_i \quad \text{and} \quad \beta = \sum_{i \in I''} a_i\varphi_i,$$

then we have that

$$\sigma \oplus \alpha = \beta.$$

Now α and β are defined over $\mathbf{Q}(\sqrt[u]{1})$, so σ is defined over $\mathbf{Q}(\sqrt[u]{1})$ by Corollary 2.4.5 (letting $R = \mathbf{Q}(\sqrt[u]{1})(G)$, $R' = \mathbf{C}(G)$ there), as claimed. $\square$

Remark. This result was originally conjectured by Schur.

4.6. Degrees of Irreducible Representations

In this section we collect, for the convenience of the reader, the various results we have proven about degrees of irreducible representations. (Of course, these results are not independent.)

We let $\mathbf{F}$ denote a splitting field for G of characteristic 0 or prime to n (for example, $\mathbf{F}$ algebraically closed). We let $\{d_i\}$ be the degrees of the irreducible $\mathbf{F}$-representations of G, and set $d_1 = 1$, the degree of the trivial representation.

Theorem 3.2.1. $d_i = 1$ for each i if and only if G is abelian.

Proposition 3.2.5. The number of i such that $d_i = 1$ is equal to the index of $[G : G]$ (the commutator subgroup of G) in G.

Theorem 3.3.3. The number of irreducible $\mathbf{F}$-representations of G is equal to t, the number of conjugacy classes of elements of G.

Theorem 3.3.3. $n = \sum_{i=1}^{t} d_i^2$.

Corollary 3.3.11. $d_i^2 \leq [G : Z(G)]$, where $Z(G)$ denotes the center of G.

Corollary 3.5.7. If $\operatorname{char}(\mathbf{F}) = p > 0$, d_i is relatively prime to p for each i.

Theorem 3.6.3. If $\operatorname{char}(\mathbf{F}) = 0$, d_i divides n for each i.

Theorem 3.6.7. $\{d_i\}$ is independent of $\operatorname{char}(\mathbf{F})$.

Theorem 4.1.14. If A is an abelian normal subgroup of G, then d_i divides $[G : A]$ for each i.

Proposition 4.1.19. For every subgroup H of G, and some irreducible representation W of H, $d_i \leq [G : H] \deg(W)$ for each i. In particular, for every abelian subgroup A of G, $d_i \leq [G : A]$ for each i.

Chapter 5

Introduction to Modular Representations

We now consider modular representation theory—the case of representations of a finite group G over a field $\mathbf{F}$ whose characteristic divides the order of G. Here the situation is considerably different, and more difficult, than the situation we have been considering thus far, though there are definite points of similarity.

We shall adopt the same approach as in the semisimple case. First we shall present some examples, and point out some salient features. Next we will develop some general ring theory to enable us to study modules over nonsemisimple rings. We shall then apply this theory to the study of $\mathbf{F}(G)$-modules, i.e., representations of G. In addition to these general results, we shall then study modular representations in particular. Throughout this development, the reader will do well to keep in mind the semisimple case in order to see both the similarities and differences between it and the modular case.

We will use the following notation henceforth.

Notation. G is a finite group and p is a prime dividing the order of G.

(1) n denotes the order of G, and $n = p^r q$ with q relatively prime to p.

(2) P denotes a p-Sylow subgroup of G.

(3) $\mathbf{F}$ denotes a field of characteristic p.

We begin with some examples.

Example 1.1. Let G be a cyclic group of order $n = p^r$ with generator a. Let $\mathbf{F}$ be a field of characteristic p. Let M be any $\mathbf{F}$-representation of G, given by $\sigma : G \to \mathrm{Aut}(M)$. Then $\alpha = \sigma(a)$ has the property that $\alpha^{p^r} = 1$. Thus α satisfies the polynomial $x^{p^r} - 1 = (x - 1)^{p^r}$, so the minimum polynomial of α divides $(x - 1)^{p^r}$, and the characteristic polynomial of α is a power of $x - 1$. Thus, α has the only eigenvalue 1, so the eigenspace of 1 is a subspace of M, i.e., M contains the 1-dimensional trivial representation τ. Hence τ is the only irreducible representation of G. Furthermore, writing the matrix of α in Jordan canonical form, we see that M is indecomposable if and only if there is a single Jordan block, which may be of any size s with $1 \leq s \leq p^r$. Thus G has a single indecomposable representation of degree s for each $1 \leq s \leq p^r$, and every indecomposable representation is one of these.

Note in particular that $\mathbf{F}(G)$ is an indecomposable representation of G of degree p^r.

Example 1.2. Let G be a p-group and let $\mathbf{F}$ be a field of characteristic p. We claim that the 1-dimensional trivial representation τ is the only irreducible representation of G, and that $\mathbf{F}(G)$ is indecomposable as an $\mathbf{F}(G)$-module. We prove the first claim by induction on n. Let $\sigma : G \to \mathrm{Aut}(M)$ be any $\mathbf{F}$-representation of G. If G is cyclic, then by Example 1.1 we are done. If not, let H be a cyclic subgroup of the center of G, generated by an element h. (Recall that, as we observed in the proof of Lemma 4.4.5, a p-group has a nontrivial center.) Let M_0 be the largest subspace of M on which H acts trivially. Again by Example 1.1, $M_0 \neq \{0\}$. As H is central, M_0 is a subrepresentation of M. (If $m_0 \in M_0$, so $\sigma(h)(m_0) = m_0$, then $\sigma(h)(\sigma(g)(m_0)) = \sigma(hg)(m_0) = \sigma(gh)(m_0) = \sigma(g)(\sigma(h)(m_0)) = \sigma(g)(m_0)$, so $\sigma(g)(m_0) \in M_0$.) But then the action of G on M_0 factors through the quotient G/H, so, by induction, τ is a subrepresentation of M_0 as a representation of G/H, and hence also as a representation of G. As for the second claim, if $\mathbf{F}(G)$ were decomposable, say $\mathbf{F}(G) = M_1 \oplus M_2$, then each of M_1 and M_2 would contain τ as a subrepresentation, so $\mathbf{F}(G)$ would contain $\tau \oplus \tau$ as a subrepresentation. But it is easy to check that the subspace of $\mathbf{F}(G)$ fixed by every element of G consists exactly of the multiples of $\sum_{g \in G} g$, which is 1-dimensional, a contradiction.

Example 1.3. Let $G = \mathbb{Z}/p\mathbb{Z} \times \mathbb{Z}/p\mathbb{Z}$ with generators a and b. Let $\mathbf{F}$ be a field of characteristic p. For each $x \in \mathbf{F}$, let $V_x = \mathbf{F}^2$, and let $\sigma_x : G \to \mathrm{Aut}(V_x)$ be defined by

$$\sigma_x(a) = \begin{bmatrix} 1 & 1 \\ 0 & 1 \end{bmatrix}, \qquad \sigma_x(b) = \begin{bmatrix} 1 & x \\ 0 & 1 \end{bmatrix}.$$

We claim that $\{V_x\}$ are distinct indecomposable representations of G.

Each V_x is clearly indecomposable. To see they are distinct, suppose $\varphi : V_x \to V_y$ is an isomorphism. After choosing bases, we may represent φ by an invertible matrix M. But it is easy to check that

$$M\sigma_x(a) = \sigma_y(a)M,$$
$$M\sigma_x(b) = \sigma_y(b)M$$

implies $x = y$.

A similar computation shows that any field over which V_x is defined must contain x.

Example 1.4 (Heller and Reiner). Let $G = \mathbb{Z}/p\mathbb{Z} \times \mathbb{Z}/p\mathbb{Z}$ with generators a and b. Let $\mathbf{F}$ be a field of characteristic p. Let M_m be the $\mathbf{F}$-vector space with basis $\{x_0, \dots, x_m, y_1, \dots, y_m\}$ and let $\sigma : G \to \operatorname{Aut}(M_m)$ be as follows:

$$\sigma(a)(x_i) = x_i, \quad \sigma(b)(x_i) = x_i,$$
$$\sigma(a)(y_i) = y_i + x_i, \quad \sigma(b)(y_i) = y_i + x_{i-1}.$$

Then M_m, of dimension $2m + 1$, is an indecomposable representation of G.

To see this, let $\alpha = \sigma(a) - 1$ and $\beta = \sigma(b) - 1$ (so $\alpha(x_i) = 0 = \beta(x_i)$, $\alpha(y_i) = x_i$, $\beta(y_i) = x_{i-1}$) and write $M_m = X \oplus Y$ (in the obvious notation), a direct sum of $\mathbf{F}$-vector spaces (*not* of representations of G). Let $\pi : M_m \to Y$ be the projection.

Let $N \neq \{0\}$ be a subrepresentation of M_m, and let $r = \dim_{\mathbf{F}} \pi(N)$. We claim that $\dim_{\mathbf{F}} N \geq 2r + 1$. This is trivially true if $r = 0$, so assume $r > 0$. Now, as N is a representation, $\alpha(N) \subseteq N$, and the restriction of α to $\operatorname{Ker}(\pi)$ is 0, so $\alpha(N) = \alpha(\pi(N))$. Since $\pi(N) \subseteq Y$, we see that the restriction of α to $\pi(N)$ is an injection, so $\dim_{\mathbf{F}} \alpha(N) = \dim_{\mathbf{F}} \alpha(\pi(N)) = \dim_{\mathbf{F}}(\pi(N)) = r$. Now pick elements $n_1, \dots, n_r \in N$ with $\{\pi(n_i)\}$ a basis for $\pi(N)$. Then $\{n_i, \alpha(n_i)\}$ is a linearly independent set (suppose $\sum c_i n_i + d_i \alpha(n_i) = 0$; applying π gives all $c_i = 0$ and then all $d_i = 0$). Also, for some i_0, $\beta(n_{i_0})$ is not in the span of the $\alpha(n_i)$ (indeed, this is true for an n_{i_0} such that the expression of $\pi(n_{i_0})$ as a linear combination of $\{y_j\}$ involves a y_j for minimal j), so $\{n_1, \dots, n_r, \alpha(n_1), \dots, \alpha(n_r), \beta(n_{i_0})\}$ is also a linearly independent set (suppose $\beta(n_{i_0}) = \sum c_i n_i + d_i \alpha(n_i)$; applying π gives all $c_i = 0$, a contradiction), proving the claim.

Now suppose $M_m = N_1 \oplus N_2$ as representations of G, and let $r_i = \dim_{\mathbf{F}} \pi(N_i)$. Note that $r_1 + r_2 \geq \dim_{\mathbf{F}} Y = m$. Then $2m + 1 = \dim_{\mathbf{F}} M_m = \dim_{\mathbf{F}} N_1 + \dim_{\mathbf{F}} N_2 \geq (2r_1 + 1) + (2r_2 + 1) = 2(r_1 + r_2) + 2 \geq 2m + 2$, a contradiction, so M_m is indecomposable.

Now we turn to a description of our agenda for the rest of this book.

Recall that a semisimple module has a unique simple factorization. We begin our general ring theory by showing that if M is an R-module "of finite length" (a condition we will define below), then the generalization of this holds, in the appropriate sense. But actually, there are two generalizations, since irreducible and indecomposable modules are (very) different. We prove each. The one in terms of irreducible modules is known as the Jordan-Holder theorem, and the one in terms of indecomposable modules is known as the Krull-Schmidt theorem.

Naturally we will want to relate our situation here to the semisimple case, so we will next study the Jacobson radical $J(R)$ of the ring R; if R has finite length, $R/J(R)$ is semisimple.

The examples we have seen indicate that irreducible (i.e., simple) R-modules may be reasonable objects of study, while indecomposable modules in full generality seem relatively intractable. We will isolate a subclass of indecomposable modules, the "principal indecomposable modules", to study. This is a natural class—indeed, these are the projective indecomposable modules—and we shall see that there is a natural one-to-one correspondence between irreducible R-modules and principal indecomposable R-modules, again when R is of finite length.

However, it will be the simple $\mathbf{F}(G)$-modules, i.e., the irreducible representations of G, that will be the focus of our attention. Here we will not concern ourselves with the question of the field of definition—to do so would require us to introduce more background—but simply assume that $\mathbf{F}$ is a large enough finite field, or perhaps $\mathbf{F} = \overline{\mathbf{F}}_p$, the algebraic closure of the field of p elements. If there are t' distinct irreducible $\mathbf{F}(G)$-modules, of degrees $d_1, \ldots, d_{t'}$, we will easily get a theoretical (though not practical, as one term is very difficult to compute) formula for $\sum d_i^2$, which in the semisimple case is n. With considerably more effort we will get a quite practical answer for t', which in the semisimple case is t, the number of conjugacy classes of elements of G. In the modular case t' is the number of conjugacy classes of p-regular elements of G. (It turns out that in the modular case the degrees d_i need not divide n.) We shall also obtain a number of other structural results difficult to summarize here.

As we have just seen in Example 1.2, in case G is a p-group, simple representations of G in characteristic p are as easy as possible. There is only one of them, τ, the 1-dimensional trivial representation. More generally, we shall see that if the p-Sylow subgroup of G is normal, the simple representations of G in characteristic p are in one-to-one correspondence with those of G/P. Note that G/P has order prime to p, so we are back in the semisimple case. Of course, for general G the subgroup P is not normal, and the situation is much more involved.

In the semisimple case, one of our main computational tools was characters. Here we must come up with a substitute, and we do, developing the "Brauer characters".

Finally, we should remind the reader that in our development of induced representations in Sections 4.1 and 4.2 we were careful to explicitly state any restrictions on G and its subgroups. Many of our results there had no restrictions, so we will be able to apply induced representations to the modular case, where they remain an important tool.

Chapter 6

General Rings and Modules

6.1. Jordan-Holder and Krull-Schmidt Theorems

We adopt the convention in this section that if ℓ is finite, $\ell < \infty$ and $\ell + \infty = \infty$, and also that $\infty + \infty = \infty$.

Definition 1.1.

(1) Let R be a ring and let M be an R-module. A **chain of submodules** of M is a finite sequence $\{M_i\}_{i=0}^{k}$ of submodules of M such that

$$(1.1) \qquad \{0\} = M_0 \subset M_1 \subset M_2 \subset \cdots \subset M_k = M.$$

The **length** of the chain is k.

(2) A chain $\{N_j\}_{j=0}^{m}$ is a **refinement** of the chain $\{M_i\}_{i=0}^{k}$ if each M_i is equal to N_j for some j. Refinement of chains defines a partial order on the set $\mathcal{C}$ of all chains of submodules of M.

(3) A maximal element of $\mathcal{C}$ (if it exists) is a **composition series** of M.

Remark 1.2.

(1) Note that the chain (1.1) is a composition series if and only if each of the modules M_i/M_{i-1} ($1 \leq i \leq k$) is a simple R-module.

(2) If $M = \bigoplus_{i=1}^{\ell} M_i$, where each M_i is a simple R-module, then M has a composition series

$$\{0\} \subset M_1 \subset M_1 \oplus M_2 \subset \cdots \subset M_1 \oplus \cdots \oplus M_\ell = M.$$

However, if $M = \bigoplus_{i=1}^{\infty} M_i$, then M does not have a composition series.

Example 1.3.

(1) Let D be a division ring and let M be a D-module with basis $\{x_1, \ldots, x_\ell\}$. Let $M_0 = \{0\}$ and for $1 \le i \le \ell$ let $M_i = \langle x_1, \ldots, x_i \rangle$. Then $\{M_i\}_{i=0}^{\ell}$ is a chain of submodules of length ℓ, and since

$$M_i/M_{i-1} = \langle x_1, \ldots, x_i \rangle / \langle x_1, \ldots, x_{i-1} \rangle$$
$$\cong Dx_i$$
$$\cong D,$$

we conclude that this chain is a composition series because D is a simple D-module.

(2) If p is a prime, the chain

$$\{0\} \subset p\mathbb{Z}/p^2\mathbb{Z} \subset \mathbb{Z}/p^2\mathbb{Z}$$

is a composition series for the $\mathbb{Z}$-module $\mathbb{Z}/p^2\mathbb{Z}$. Note that $\mathbb{Z}/p^2\mathbb{Z}$ is not semisimple as a $\mathbb{Z}$-module as $p\mathbb{Z}/p^2\mathbb{Z}$ is not a direct summand.

(3) The $\mathbb{Z}$-module $\mathbb{Z}$ does not have a composition series. $\{0\} \subset \mathbb{Z}$ is certainly not a composition series. Furthermore, if $\{I_i\}_{i=0}^{\ell}$ is any chain of submodules of length $\ell > 1$, then writing $I_1 = \langle a_1 \rangle$, we can properly refine the chain by putting the ideal $\langle a_1^2 \rangle$ between I_1 and $I_0 = \{0\}$.

(4) If R is a ring without zero divisors that is not a division ring, then a similar argument shows that R does not have a composition series as an R-module.

Definition 1.4. Let M be an R-module. If M has a composition series, let $\ell(M)$ denote the minimum length of a composition series for M. If M does not have a composition series, let $\ell(M) = \infty$. $\ell(M)$ is called the **length** of the R-module M. If $\ell(M) < \infty$, we say that M has **finite length**.

Note that isomorphic R-modules have the same length, since if $f : M \to N$ is an R-module isomorphism, the image under f of a composition series for M is a composition series for N.

Note that if M is a finite-dimensional vector space over a field $\mathbf{F}$, its length as an $\mathbf{F}$-module is equal to its dimension as a $\mathbf{F}$-vector space, while if M is infinite dimensional, its length is ∞. The reader should keep in mind the very strong analogy between the length of a module and the dimension of a vector space in results 1.5 through 1.12 below.

Lemma 1.5. *Let M be an R-module of finite length and let $N \subset M$ be a proper submodule. Then $\ell(N) < \ell(M)$.*

Proof. Let

$$\{0\} = M_0 \subset M_1 \subset \cdots \subset M_\ell = M \tag{1.2}$$

be a composition series of M of length $\ell = \ell(M)$ and let $N_i = N \cap M_i \subseteq N$. Let $\varphi : N_i \to M_i/M_{i-1}$ be the inclusion $N_i \to M_i$ followed by the projection $M_i \to M_i/M_{i-1}$. Since $\mathrm{Ker}(\varphi) = N_{i-1}$, it follows that N_i/N_{i-1} is isomorphic to a submodule of M_i/M_{i-1}. But M_i/M_{i-1} is a simple R-module, as (1.2) is a composition series. Hence $N_i = N_{i-1}$ or $N_i/N_{i-1} = M_i/M_{i-1}$ for $i = 1, 2, \ldots, \ell$. By deleting the repeated terms of the sequence $\{N_i\}_{i=0}^\ell$ we obtain a composition series for the module N of length $\leq \ell = \ell(M)$. Suppose that this composition series for N has length ℓ. Then we must have $N_i/N_{i-1} = M_i/M_{i-1}$ for all $i = 1, 2, \ldots, \ell$. Thus $N_1 = M_1$, $N_2 = M_2$, $\ldots, N_\ell = M_\ell$, i.e., $N = M$. The lemma follows by contraposition. $\qquad\square$

Proposition 1.6. *Let M be an R-module of finite length. Then every composition series of M has length $\ell = \ell(M)$. Moreover, every chain of submodules can be refined to a composition series.*

Proof. We first show that any chain of submodules of M has length $\leq \ell(M)$. Let

$$\{0\} = M_0 \subset M_1 \subset \cdots \subset M_k = M$$

be a chain of submodules of M of length k. By Lemma 1.5,

$$\{0\} = \ell(M_0) < \ell(M_1) < \cdots < \ell(M_k) = \ell(M).$$

Thus, $k \leq \ell(M)$.

Now consider a composition series of M of length k. By the definition of composition series, $k \geq \ell(M)$, and we just proved that $k \leq \ell(M)$. Thus, $k = \ell(M)$. If a chain has length $\ell(M)$, then it must be maximal and, hence, is a composition series. If the chain has length $< \ell(M)$, then it is not a composition series and hence it may be refined until its length is $\ell(M)$, at which time it will be a composition series. $\qquad\square$

Proposition 1.7. *Let D be a division ring and let M be a D-module with a finite basis. Then every basis of M is finite and all bases have the same number of elements.*

Proof. By Example 1.3 (1), if D is a division ring and M is a D-module, then a basis $S = \{x_1, \ldots, x_\ell\}$ with ℓ elements determines a composition series of M of length ℓ. Since all composition series of M must have the same length, we conclude that any two finite bases of M must have the same length ℓ. Moreover, if M also had an infinite basis T, then M would have a linearly independent set consisting of more than ℓ elements. Call this set $\{y_1, \ldots, y_k\}$ with $k > \ell$. Then

$$\{0\} \subset \langle y_1 \rangle \subset \langle y_1, y_2 \rangle \subset \cdots \subset \langle y_1, \ldots, y_k \rangle \subseteq M$$

is a chain of length $> \ell$, which contradicts Proposition 1.6. Thus, every basis of M is finite and has ℓ elements. $\square$

An (almost) equivalent way to state the same result is the following. It can be made equivalent by the convention that D^∞ refers to any infinite direct sum of copies of D, without regard to the cardinality of the index set.

Corollary 1.8. *If D is a division ring and $D^m \cong D^\ell$, then $m = \ell$.*

Proof. $\square$

Remark. It is not true for an arbitrary ring R that $R^m \cong R^\ell$ implies $m = \ell$. For example, let $R = \{f : \mathbb{Z} \to \mathbb{Z}\}$. Then $R = R_{\mathrm{even}} \oplus R_{\mathrm{odd}}$ where $R_{\mathrm{even}} = \{f : \{\text{even integers}\} \to \mathbb{Z}\}$ and $R_{\mathrm{odd}} = \{f : \{\text{odd integers}\} \to \mathbb{Z}\}$, and clearly $R_{\mathrm{even}} \cong R$, $R_{\mathrm{odd}} \cong R$, so $R \cong R^2$.

Proposition 1.9. *Let* $0 \longrightarrow K \overset{\varphi}{\longrightarrow} M \overset{\psi}{\longrightarrow} L \longrightarrow 0$ *be a short exact sequence of R-modules. Then*

$$\ell(M) = \ell(K) + \ell(L).$$

Proof. We prove the most interesting case. Let

$$\{0\} = K_0 \subset K_1 \subset \cdots \subset K_k = K$$

be a composition series for K, and let

$$\{0\} = L_0 \subset L_1 \subset \cdots \subset L_m = L$$

be a composition series for L. For $0 \leq i \leq k$, let $M_i = \varphi(K_i)$, and for $k + 1 \leq i \leq k + m$, let $M_i = \psi^{-1}(L_{i-k})$. Then $\{M_i\}_{i=0}^{k+m}$ is a chain of submodules of M and

$$M_i/M_{i-1} \cong \begin{cases} K_i/K_{i-1} & \text{for } 1 \leq i \leq k, \\ L_{i-k}/L_{i-k-1} & \text{for } k + 1 \leq i \leq k + m, \end{cases}$$

so that $\{M_i\}_{i=0}^{k+m}$ is a composition series for M. Thus, $\ell(M) = k + m$. $\square$

Corollary 1.10.

(1) *Let K and L be submodules of M. Then*

$$\ell(K + L) + \ell(K \cap L) = \ell(K) + \ell(L).$$

(2) *Let $\ell(M)$ be finite. If $\ell(K) + \ell(L) < \ell(M)$, then $K + L \neq M$. If $\ell(K) + \ell(L) > \ell(M)$, then $K \cap L \neq \{0\}$. If $\ell(K) + \ell(L) = \ell(M)$, then $K + L = M$ if and only if $K \cap L = \{0\}$.*

Proof.

(1) By Proposition 1.9, we have that $\ell(K + L) = \ell(K) + \ell((K+L)/K)$. But $(K + L)/K$ is isomorphic to $L/L \cap K$, so $\ell((K + L)/K) = \ell(L/L \cap K) = \ell(L) - \ell(L \cap K)$, again by Proposition 1.9.

(2) This follows immediately from (1) and Lemma 1.5. $\square$

Lemma 1.11. *Let* $f : M \to N$ *be an* R-*module homomorphism.*

(1) *If* f *is an injection, then* $\ell(M) \le \ell(N)$.

(2) *If* f *is a surjection, then* $\ell(M) \ge \ell(N)$.

(3) *If* $\ell(M) = \ell(N) < \infty$, *the following are equivalent:*

    (i) f *is an injection.*

    (ii) f *is a surjection.*

    (iii) f *is an isomorphism.*

Proof. Straightforward. $\qquad\qquad\qquad\qquad\qquad\qquad\qquad\qquad\qquad\square$

Corollary 1.12. *Let* M *be an* R-*module of finite length. Then* M *is not isomorphic to any proper submodule of* M *or to the quotient of* M *by any nontrivial submodule of* M.

Proof. $\qquad\qquad\qquad\qquad\qquad\qquad\qquad\qquad\qquad\qquad\qquad\qquad\qquad\square$

Lemma 1.13. *Let* M *be an* R-*module of finite length, and let* $f \in \mathrm{End}_R(M)$. *Then there exists a* $k_0 \le \ell(M)$ *such that*

$$M \cong \mathrm{Ker}(f^k) \oplus \mathrm{Im}(f^k)$$

for every $k \ge k_0$.

Proof. Let $\ell = \ell(M)$ and consider

$$\{0\} \subseteq f^\ell(M) \subseteq f^{\ell-1}(M) \subseteq \cdots \subseteq f(M) \subseteq f^0(M) = M.$$

This cannot be a chain of submodules as its length is $\ell + 1$. Thus we must have equality at some point in the sequence, so let k_0 be the smallest exponent for which $f : f^{k_0}(M) \to f^{k_0+1}(M)$ is an isomorphism for $k_0 < \ell$, or $k_0 = \ell$ otherwise (in which case $f^\ell(M) = \{0\}$). Let $k \ge k_0$.

Let $M' = f^k(M)$ and note that $f \mid M' : M' \to M'$ is an isomorphism by Lemma 1.11. Let $\varphi = f^k \mid M' : M' \to M'$, also an isomorphism.

If $m' \in M'$ with $\varphi(m') = 0$, then $m' = 0$, so $\mathrm{Ker}(f^k) \cap \mathrm{Im}(f^k) = \{0\}$.

Let $m \in M$ and let $m' = \varphi^{-1}(f^k(m))$. Then $m = (m - m') + m'$ and also $f^k(m - m') = f^k(m) - f^k(m') = f^k(m) - f^k(\varphi^{-1}(f^k(m))) = f^k(m) - f^k(m) = 0$ so $\mathrm{Ker}(f^k) + \mathrm{Im}(f^k) = M$, and we are done. $\quad\square$

Corollary 1.14. *Let* M *be an indecomposable* R-*module of finite length, and let* $f \in \mathrm{End}_R(M)$. *Then either* f *is an isomorphism or* $f^k = 0$ *for some* $k \le \ell(M)$.

Proof. $\qquad\qquad\qquad\qquad\qquad\qquad\qquad\qquad\qquad\qquad\qquad\qquad\qquad\square$

Suppose that $M = N_1 \oplus N_2$ with N_1 and N_2 simple. Then we have composition series $\{0\} \subset N_1 \subset M$ and $\{0\} \subset N_2 \subset M$, so we cannot expect composition series to be unique. The next theorem tells us that the best we can hope for is indeed true.

Theorem 1.15 (Jordan-Holder). *Let M be an R-module of finite length $\ell = \ell(M)$ and let*

$$\{0\} \subset M_1 \subset M_2 \subset \cdots \subset M_\ell = M$$

and

$$\{0\} \subset N_1 \subset N_2 \subset \cdots \subset N_\ell = M$$

be two composition series for M. Then

$$\{[M_i/M_{i-1}], \ i = 1, \ldots, \ell\} \ = \ \{[N_i/N_{i-1}], \ i = 1, \ldots, \ell\}$$

as sets with multiplicities, where $[\ ]$ denotes isomorphism class.

Proof. We prove the theorem by induction on ℓ. If $\ell = 1$, it is trivial.

Though not strictly necessary, we prove the case $\ell = 2$ separately, as this case shows the essential idea of the proof. Thus, suppose $\ell(M) = 2$ and let M have two composition series as above.

If $M_1 \subseteq N_1$, or $N_1 \subseteq M_1$, then they are equal, as M_1 and N_1 are both simple, and then the theorem is trivial.

Suppose not. Then $M_1 \subset M_1 + N_1 \subseteq M$ so $M_1 + N_1 = M$ as M/M_1 is simple. Also, $\{0\} \subseteq M_1 \cap N_1 \subset M_1$ so $\{0\} = M_1 \cap N_1$ as M_1 is simple. Then

$$M/M_1 = (M_1 + N_1)/M_1 \cong N_1/M_1 \cap N_1 = N_1/\{0\} \cong N_1$$

and

$$M/N_1 = (M_1 + N_1)/N_1 \cong M_1/M_1 \cap N_1 = M_1/\{0\} \cong M_1.$$

Thus

$$\{[M_1], [M/M_1]\} \ = \ \{[M_1], [N_1]\} \ = \ \{[N_1], [M_1]\} \ = \ \{[N_1], [M/N_1]\}$$

as claimed.

Now for the general inductive case. Suppose the theorem is true for all R-modules of length $\ell - 1$, and consider an R-module of length ℓ with two composition series as above.

If $M_{\ell-1} \subseteq N_{\ell-1}$, or $N_{\ell-1} \subseteq M_{\ell-1}$, then they are equal by Lemma 1.5 as both have the same length $\ell - 1$. Then

$$\{[M_i/M_{i-1}], \ i = 1,\ldots,\ell - 1\} = \{[N_i/N_{i-1}], \ i = 1,\ldots,\ell - 1\}$$

by induction, and $M/M_{\ell-1} = M/N_{\ell-1}$, so we are done.

Otherwise $M_{\ell-1} + N_{\ell-1} = M$, as $M/M_{\ell-1}$ is simple. Also, let $L = M_{\ell-1} \cap N_{\ell-1}$ and note by Corollary 1.10 (1) that $\ell(L) = \ell - 2$. Write $L = L_{\ell-2}$. Then

$$M/M_{\ell-1} = (M_{\ell-1} + N_{\ell-1})/M_{\ell-1} \cong N_{\ell-1}/N_{\ell-1} \cap M_{\ell-1} = N_{\ell-1}/L_{\ell-2}$$

and

$$M/N_{\ell-1} = (N_{\ell-1} + M_{\ell-1})/N_{\ell-1} \cong M_{\ell-1}/M_{\ell-1} \cap N_{\ell-1} = M_{\ell-1}/L_{\ell-2},$$

and each of these quotients is simple.

Let

$$\{0\} \subset L_1 \subset \cdots \subset L_{\ell-2} = L$$

be a composition series for L. We then have composition series

$$\{0\} \subset L_1 \subset \cdots \subset L_{\ell-2} \subset M_{\ell-1} \subset M$$

and

$$\{0\} \subset L_1 \subset \cdots \subset L_{\ell-2} \subset N_{\ell-1} \subset M.$$

We now have two composition series for both $M_{\ell-1}$ and $N_{\ell-1}$, so by induction we may conclude that

$$\{[L_i/L_{i-1}], \ i = 1,\ldots,\ell - 2, [M_{\ell-1}/L_{\ell-2}]\} = \{[M_i/M_{i-1}], \ i = 1,\ldots,\ell - 1\}$$

and

$$\{[L_i/L_{i-1}], \ i = 1,\ldots,\ell - 2, [N_{\ell-1}/L_{\ell-2}]\} = \{[N_i/N_{i-1}], \ i = 1,\ldots,\ell - 1\}.$$

Then

$$\begin{aligned}
\{[M_i/M_{i-1}], i = 1,\ldots,\ell\} &= \{[M_i/M_{i-1}], i = 1,\ldots,\ell - 1\} \cup \{[M/M_{\ell-1}]\} \\
&= \{[M_i/M_{i-1}], i = 1,\ldots,\ell - 1\} \cup \{[N_{\ell-1}/L_{\ell-2}]\} \\
&= \{[L_i/L_{i-1}], i = 1,\ldots,\ell - 2, [M_{\ell-1}/L_{\ell-2}]\} \\
&\qquad\qquad \cup \{[N_{\ell-1}/L_{\ell-2}]\} \\
&= \{[L_i/L_{i-1}], i = 1,\ldots,\ell - 2, [N_{\ell-1}/L_{\ell-2}]\} \\
&\qquad\qquad \cup \{[M_{\ell-1}/L_{\ell-2}]\} \\
&= \{[N_i/N_{i-1}], i = 1,\ldots,\ell - 1\} \cup \{[M_{\ell-1}/L_{\ell-2}]\} \\
&= \{[N_i/N_{i-1}], i = 1,\ldots,\ell - 1\} \cup \{[M/N_{\ell-1}]\} \\
&= \{[N_i/N_{i-1}], i = 1,\ldots,\ell\}
\end{aligned}$$

as claimed. $\qquad\qquad\qquad\qquad\qquad\qquad\qquad\qquad\qquad\qquad\qquad\qquad \square$

Definition 1.16. If M is an R-module of finite length $\ell = \ell(M)$, and

$$\{0\} \subset M_1 \subset \cdots \subset M_\ell = M$$

is a composition series for M, then the type $\mathcal{T}(M)$ of M is

$$\mathcal{T}(M) = \{[M_i/M_{i-1}], \ i = 1, \ldots, \ell\},$$

a set with multiplicities.

Note that our definitions of length and type there agree with those in Section 2.2 for semisimple R-modules.

Now we turn to the consideration of indecomposable modules.

Lemma 1.17. *Let M be an indecomposable R-module and N an arbitrary nonzero R-module. Let $\alpha \in \mathrm{Hom}_R(N, M)$ and $\beta \in \mathrm{Hom}_R(M, N)$ be such that $\beta\alpha : N \to N$ is an isomorphism. Then α and β are each isomorphisms.*
Proof. Let $K = \mathrm{Ker}(\beta)$ and $I = \mathrm{Im}(\alpha)$. We show $M = K \oplus I$; since M is indecomposable, the lemma follows.

For any $m \in M$, $m - \alpha((\beta\alpha)^{-1}(\beta(M))) \in \mathrm{Ker}(\beta)$, so $K + I = M$.

On the other hand, let $m \in K \cap I$. Then $m = \alpha(m')$ and $0 = \beta(m) = \beta\alpha(m')$ implies $m' = 0$, so $K \cap I = \{0\}$. $\square$

Lemma 1.18. *Let M be an indecomposable R-module of finite length, and let $\theta_i \in \mathrm{End}_R(M), i = 1, \ldots, r$, with $f = \sum \theta_i : M \to M$ an isomorphism. Then some θ_i is an isomorphism.*
Proof. We argue by induction on r. If $r = 1$ there is nothing to prove.

Suppose that $r = 2$. Let $\varphi_i = f^{-1}\theta_i$ so $\varphi_1 + \varphi_2 = 1_M$. Then φ_1 and φ_2 commute, as $\varphi_1\varphi_2 = \varphi_1(1_M - \varphi_1) = (1_M - \varphi_1)\varphi_1 = \varphi_2\varphi_1$. Suppose that neither φ_1 nor φ_2 is an isomorphism. Then corollary 1.14 gives that $\varphi_1^{k_1} = 0$ and $\varphi_2^{k_2} = 0$ for some k_1 and k_2. Then, setting $k = k_1 + k_2$, $1_M = (1_M)^k = (\varphi_1 + \varphi_2)^k = 0$ (by the binomial theorem), a contradiction.

Suppose that $r > 2$. Write $\theta_1 + \cdots + \theta_r = \theta_1 + \theta_2'$, where $\theta_2' = \theta_2 + \cdots \oplus \theta_r$. By the $r = 2$ case, either θ_1 is an isomorphism, and we are done immediately, or θ_2' is an isomorphism, in which case (replacing f by θ_2') we are again done by induction. $\square$

Theorem 1.19. *Let M be an R-module of finite length. Then:*

(1) *M can be written as a direct sum of finitely many indecomposable R-modules.*

(2) (**Krull-Schmidt**) *If*

$$M = M_1 \oplus \cdots \oplus M_k = N_1 \oplus \cdots \oplus N_\ell,$$

then $k = \ell$ and, after possible reordering, M_i and N_i are isomorphic for each i.

Proof. First we show existence (i.e., (1)) and then essential uniqueness (i.e., (2)) of such a direct sum.

For existence, we argue by induction on the length $\ell(M)$ of M. If $\ell(M) = 1$, then M is indecomposable, and we are done. Let $\ell(M) > 1$. Either M is indecomposable, and we are done, or $M = M' \oplus M''$ for proper submodules M' and M''. But then, by Lemma 1.5, $\ell(M') < \ell(M)$, $\ell(M'') < \ell(M)$, so each of M' and M'' is a sum of indecomposable R-modules, and hence so is M.

Now for essential uniqueness. We may assume $k \leq \ell$, and we prove the claim by induction on k. If $k = 1$, M is indecomposable, and we are done.

Let μ_i be the projection $\mu_i : M \to M_i$ and let ν_j be the projection $\nu_j : M \to N_j$. Then $\mu_1 = \mu_i 1 = \mu_1(\nu_1 + \cdots + \nu_j) = \mu_1\nu_1 + \cdots + \mu_1\nu_j$. Restricted to M_1, μ_1 is the identity, so by Lemma 1.18 some $\mu_1\nu_j$, say $\mu_1\nu_1$, is an isomorphism $\mu_1\nu_1 : M_1 \to M_1$. By Lemma 1.17, this implies that $\nu_1 : M_1 \to N_1$ and $\mu_1 : N_1 \to M_1$ are both isomorphisms.

We claim $M = N_1 \oplus M_2 \oplus \cdots \oplus M_k$. First, $N_1 \cap (M_2 \oplus \cdots \oplus M_k) = \{0\}$ as if $n_1 = m_2 + \cdots + m_k$, $n_1 \in N_1$, $m_i \in M_i$, then $\mu_1(n_1) = \mu_1(m_2 + \cdots + m_k) = 0$, so $n_1 = 0$ as μ_1 is an isomorphism. Let $M' = N_1 \oplus M_2 \oplus \cdots \oplus M_k$. Then $\rho = \nu_1 + \mu_2 + \cdots + \mu_k$ is an isomorphism from M to $M' \subseteq M$, so by Corollary 1.12 we have $M' = M$.

Thus $\rho : M \to M$ is an isomorphism with $\rho(M_1) = N_1$, so ρ induces an isomorphism $\overline{\rho} : M/M_1 \to M/N_1$, i.e., an isomorphism $\overline{\rho} : M_2 \oplus \cdots \oplus M_k \to N_2 \oplus \cdots \oplus N_\ell$, and by induction we are done. $\square$

Remark. Our exposition here follows Curtis and Reiner [CR, Theorem 14.5].

Remark 1.20. Of course, if M is not of finite length, the existence part of Theorem 1.19 may break down. In this case, however, even if existence holds, the uniqueness part may break down. Here is an example. Let $R = \{f : [0,1] \to \mathbf{R} \mid f \text{ continuous}, f(1) = f(0)\}$. Let $M = \{f : [0,1] \to \mathbf{R} \mid f \text{ continuous}, f(1) = -f(0)\}$. Then R is a ring, M is an R-module and M and R are not isomorphic as R-modules. (R is free on the constant function 1 as an R-module, but M cannot be generated by a single element as any function in M must be zero somewhere.) However, $M \oplus M$ and $R \oplus R$ are isomorphic as R-modules. (The isomorphism is given by $\varphi(f,g)(x) = r_{\pi x}(f(x), g(x))$ where r_θ is rotation through angle θ.)

6.2. The Jacobson Radical

Definition 2.1. Let R be a ring. The **Jacobson radical** $J = J(R)$ is the intersection of all the maximal left ideals of R.

Proposition 2.2. *The Jacobson radical $J(R)$ is given by*

$$J(R) = \bigcap \operatorname{Ann}(M),$$

where the intersection is taken over all simple left R-modules M.

Proof. Suppose y annihilates every simple left R-module. For any maximal left ideal I, R/I is a simple left R-module, so $y(R/I) = 0$, so $y \in I$, and so $y \in J(R)$.

Conversely, suppose $y \in J(R)$ and let M be any simple left R-module. Then M is cyclic, generated by any nonzero element $m \in M$, and then $M \cong R/\operatorname{Ann}(m)$. Since M is simple, $\operatorname{Ann}(m)$ is maximal, so $y \in \operatorname{Ann}(m)$, and so $y \in \bigcap_{m \in M} \operatorname{Ann}(m) = \operatorname{Ann}(M)$. $\qquad\square$

Corollary 2.3. $J(R)$ *is a two-sided ideal in R.*

Proof. For any R-module M, $\operatorname{Ann}(M)$ is a two-sided ideal in R. $\qquad\square$

Proposition 2.4. *For any $y \in J(R)$, $1 + y$ is a unit in R. Indeed, $J(R)$ is the largest left ideal (and hence the largest two-sided ideal) I such that $1 + I = \{1 + y \mid y \in I\}$ is contained in $\{$units of $R\}$.*

Proof. Let $y \in J(R)$. First observe that $R(1 + y) = R$. For if not, $R(1 + y)$ would be a proper left ideal of R, so contained in some maximal left ideal M. Then $1 + y \in M$, and $y \in M$ (as $J(R) \subseteq M$) implies $1 \in M$, a contradiction. Hence $1 + y$ has a left inverse u. Thus $u(1 + y) = u + uy = 1$, $u = 1 - uy$. But $-uy \in J(R)$, so $u = 1 + (-uy)$ has a left inverse v. Hence $vu = u(1 + y) = 1$, so u is invertible, and $u^{-1} = 1 + y$, so $1 + y$ is invertible.

Conversely, let I be a left ideal of R with the property that $1 + y$ is a unit for every $y \in I$. We claim $I \subseteq J(R)$. Suppose not, and let $y \in I$, $y \notin J(R)$. Then there is some simple R-module M with $yM \neq \{0\}$. Choose any $m \in M$ with $ym \neq 0$. Then $Rym \subseteq RM = M$, so $Rym = M$ as M is simple. Hence there is an $x \in R$ with $xym = -m$, i.e., $(1 + xy)m = 0$. But $y \in I$ and I is a left ideal, so $xy \in I$ and then $1 + xy$ is a unit by the definition of I, a contradiction. $\qquad\square$

Remark 2.5. We really should have defined J to be the left Jacobson radical, and defined the right Jacobson radical similarly, but the above corollary shows that the two agree.

The following is a generally useful result.

Lemma 2.6 (Nakayama). *For any left ideal $I \subset R$, the following are equivalent:*

(1) $I \subseteq J(R)$.

(2) *For any nonzero finitely generated left R-module M, IM is a proper submodule of M.*

(3) *For any left R-modules M and N with N a proper submodule of M and M/N finitely generated, $N + IM$ is a proper submodule of M.*

Proof. $(2) \Rightarrow (1)$. If M is simple, then M is cyclic (i.e., generated by a single element of M). Also, if M is simple, its only proper submodule is $\{0\}$. Thus as a special case of (2), we have that $IM = \{0\}$ for every simple R-module M, and then $I \subseteq J(R)$ by Proposition 2.2.

$(3) \Rightarrow (2)$. Set $N = \{0\}$ in (3).

$(1) \Rightarrow (3)$. As M/N is finitely generated, it has a finite generating set, and hence a minimal generating set $\{m_1, \ldots, m_k\}$. Let N' be the proper submodule of M spanned by N and $\{m_1, \ldots, m_{k-1}\}$. Then M/N' is cyclic with generator m_k, so M/N' is isomorphic to R/K for some left ideal K. But then K is contained in a maximal left ideal K', and $J(R) \subseteq K'$ by definition, so $K + J(R) \subseteq K' \subset R$ and hence $N + J(R)M \subseteq N' + K'M$ is properly contained in M. $\qquad\square$

Proposition 2.7. *Let I be any two-sided ideal of R contained in $J(R)$. Then*
$$J(R/I) = J(R)/I.$$

Proof. $\qquad\qquad\qquad\qquad\qquad\qquad\qquad\qquad\qquad\qquad\qquad\qquad\square$

Definition 2.8. A ring R is **J-semisimple** if $J(R) = \{0\}$.

Thus in light of Proposition 2.7, $R/J(R)$ is always J-semisimple.

Proposition 2.9. *If M is a simple R-module, then $J(R)M = \{0\}$. Hence R and $R/J(R)$ have the same simple modules.*

Proof. If M is simple, $J(R) \subseteq \mathrm{Ann}(M)$. $\qquad\qquad\qquad\qquad\qquad\square$

Proposition 2.10. *Every semisimple ring is J-semisimple.*

Proof. We prove the contrapositive: If $J(R) \neq \{0\}$, then R is not semisimple. Now $J(R)$ is a left ideal of R. We show $J(R)$ is not a summand of R.

Suppose that $R = J \oplus K$ where K is a left ideal of R. Then we may write (uniquely) $1 = -j + k$, where $j \in J$, $k \in K$. Then $k = 1 + j \in K$. But $1 + j$ is a unit by Proposition 2.4, a contradiction. $\qquad\square$

Observe that $\mathbb{Z}$ is a ring that is J-semisimple but not semisimple, so the converse of this proposition does not hold. We shall return to this point in Corollary 3.11 below.

Definition 2.11.

(1) An element $r \in R$ is nilpotent if $r^k = 0$ for some k.

(2) A (left or two-sided) ideal $I \subset R$ is nil if every $r \in I$ is nilpotent.

(3) A (left or two-sided) ideal $I \subset R$ is nilpotent if $I^k = 0$ for some k (i.e., if $r_1 r_2 \cdots r_k = 0$ for every $r_1, r_2, \ldots, r_k \in I$).

Lemma 2.12. *If $r \in R$ is nilpotent, then $1 + r$ and $1 - r$ are units of R.*

Proof. Suppose $r^k = 0$. Then $1 = 1 - r^k = (1 - r)(1 + r + \cdots + r^{k-1}) = (1 + r + \cdots + r^{k-1})(1 - r)$, and similarly for $1 + r$. $\square$

Note that if a is nilpotent, it is not necessarily the case that the ideal Ra is nil. Also, if a and b are nilpotent, it is not necessarily the case that $a + b$ is nilpotent, or that ab is nilpotent. Examples for all of these claims may be obtained by letting R be the ring of two-by-two matrices (with entries in $\mathbb{Z}$, say) and $a = \left[\begin{smallmatrix} 0 & 1 \\ 0 & 0 \end{smallmatrix}\right]$, $b = \left[\begin{smallmatrix} 0 & 0 \\ 1 & 0 \end{smallmatrix}\right]$. Also, a nil ideal need not be nilpotent. For example, let $R = \mathbb{Z}[x_1, x_2, x_3, \ldots]/(x_1^2, x_2^2, x_3^2, \ldots)$ and let $\overline{x}_i$ be the image of x_i in R. Then the ideal I generated by $\overline{x}_1, \overline{x}_2, \overline{x}_3, \ldots$ is nil but not nilpotent.

Lemma 2.13. *Let $\{I_i\}_{i=1,\ldots,k}$ be a finite collection of nilpotent left ideals in R. Then $I = I_1 + \cdots + I_k$ is nilpotent.*

Proof. It suffices to prove this in case $k = 2$. Suppose $I_1^{k_1} = 0$ and $I_2^{k_2} = 0$. If $k = k_1 + k_2$, we claim $I^k = 0$. For any element of I^k can be written as $(a_1 + b_1)(a_2 + b_2) \cdots (a_k + b_k)$ where $a_i \in I_1$ and $b_i \in I_2$. Then any term in the product has at least k_1 a's or at least k_2 b's, and so is zero. (For example, if $k_2 = 2$, $a_1 b_2 a_3 b_4 = (a_1 b_2)(a_3 b_4) = 0$ as $a_1 b_2 \in I_2$ and $a_3 b_4 \in I_2$, as I_2 is a left ideal.) $\square$

Lemma 2.14. *Let I be a nil (left or two-sided) ideal. Then $I \subseteq J(R)$.*

Proof. Let $r \in I$. By definition, r is nilpotent, so, by Lemma 2.12, $1 + r$ is invertible. Then, by proposition 2.4, $I \subseteq J(R)$. $\square$

6.3. Rings of Finite Length

In this section we gather together a number of different results on rings R of finite length. Logically speaking, it may be viewed as a collection of subsections: First we discuss the Jacobson radical $J(R)$ in this case. Next we prove a basic structure theorem for R (Theorem 3.9) and derive its consequences. Next we show that if R is J-semisimple and of finite length, then R is semisimple (and conversely). Then we investigate block decompositions of R. Finally, we study the relationship between chain conditions and the condition of being of finite length.

Definition 3.1. A ring R is of **finite length** if R has finite length as a (left) R-module.

Note that $\mathbf{F}(G)$ has finite length, as it has finite dimension as an **F**-vector space. We will be applying the results of this section in the case $R = \mathbf{F}(G)$, but we proceed here more generally.

We begin by refining the theory of the Jacobson radical.

Proposition 3.2. *Let R be a ring of finite length. Then $J(R)$ is the largest nilpotent left ideal in R.*

Proof. Lemma 2.14 shows that $J(R)$ contains any nil, and hence any nilpotent, left ideal. We show $J = J(R)$ is nilpotent. Consider

$$\cdots \subseteq J^3 \subseteq J^2 \subseteq J \subseteq R.$$

Since R has finite length, we must have $J^k = J^{k+1} = \cdots$ for some k. Let $J^k = I$. We wish to show $I = \{0\}$. Suppose not. Then I has finite length, so contains a minimal nonzero ideal I_0. Choose $a \in I_0$, $a \neq 0$. Then $I_0 a \subseteq I_0$, so $I_0 a = I_0$. In particular, $ya = -a$ for some $y \in I_0$, i.e., $(1 + y)a = 0$. But $y \in I_0 \subseteq J(R)$, so, by Proposition 2.4, $1 + y$ is a unit, a contradiction. $\square$

Corollary 3.3. *Let R be a ring of finite length. Then every nil left ideal is nilpotent.*

Proof. If I is a nil left ideal, then $I \subseteq J(R)$ by Lemma 2.14, and $J(R)$ is nilpotent by Proposition 3.2, so I is nilpotent. $\square$

Corollary 3.4. *Let R be a ring of finite length and let $\{I_i\}$ be any collection of nilpotent left ideals. Then I, the sum of the I_i, is nilpotent.*

Proof. Since any element of I is in the sum of finitely many of the I_i, an argument as in the proof of Lemma 2.13 shows that I is nil, and hence is nilpotent by Corollary 3.3. $\square$

Proposition 3.5. *Let R be a ring of finite length. Then R has a maximal nilpotent left ideal W, which is the sum of all the nilpotent left ideals. This ideal W is the Jacobson radical. Also, W is a two-sided ideal.*

Proof. The first statement follows from Corollary 3.4, the second from Proposition 3.2, and the third from Corollary 2.3. $\square$

Remark 3.6. Originally the radical of R was first defined in the case where R is of finite length. This definition, given by Wedderburn, is that of W in the proposition above. Thus this proposition shows that the Jacobson radical is a generalization of the Wedderburn radical.

Definition 3.7. Let R be a ring of finite length, and write

$$R = \bigoplus S_i$$

with each S_i an indecomposable R-module. An R-module M is a **principal indecomposable** R-module if it is isomorphic to S_i for some i.

Note that the S_i are well defined up to isomorphism by the Krull-Schmidt theorem (Theorem 1.19 (2)).

Lemma 3.8. *Let R be a ring of finite length. An R-module M is projective if and only if it is isomorphic to a direct sum of principal indecomposable R-modules.*

Proof. Recall that an R-module M is projective if and only if it is a summand of a free R-module. Then the if part of the lemma is immediate, and the only-if part follows directly from the Krull-Schmidt theorem. $\square$

Theorem 3.9. *Let R be a ring of finite length. Write $R = \bigoplus S_i$ as in Definition 3.7. Then*

(1) *Each S_i contains a unique maximal proper left R-ideal J_i. This ideal J_i is $S_i \cap J(R) = J(R)S_i$.*

(2) *S_i and S_j are isomorphic if and only if $T_i = S_i/J_i$ and $T_j = S_j/J_j$ are isomorphic.*

(3) *Each T_i is simple and every simple R-module M is isomorphic to some T_i.*

Proof.

(1) Let $\pi : R \to S_1$ be the projection. Recall that $e_1 = \pi(1)$ is an idempotent (as $e_1^2 = \pi(1)\pi(1) = \pi(1 \cdot 1) = \pi(1) = e_1$) and $S_1 = Re_1$. In particular, S_1 is not a nil ideal.

Since R has finite length, S_1 has finite length by Lemma 1.5. Let A be any proper left R-ideal in S_1. We claim A is nil. For let $a \in A$, and define $\alpha : S_1 \to S_1$ by $\alpha(r) = ra$. Since $\alpha(S_1) \subseteq A \subset S_1$, α is not an isomorphism, so $\alpha^k = 0$ for some k by Corollary 1.14. Hence $ra^k = 0$ for every $r \in S_1$, so in particular $a^{k+1} = 0$.

Since A is nil, A is contained in $J(R)$ by Lemma 2.14. Then $S_1 \cap J(R)$ is the unique maximal left R-ideal in S_1 and is proper as $J(R)$ is nil and S_1 is not.

Finally, $J(R)S_1$ is contained in $S_1 \cap J(R)$ as S_1 is a left ideal and $J(R)$ is a two-sided ideal. On the other hand, if $x \in S_1 \cap J(R)$, then $x = re_1 = re_1^2 = (re_1)e_1 = xe_1 \in J(R)S_1$.

(2) Clearly any isomorphism $f : S_1 \to S_2$ has $f(J_1) = J_2$ and hence induces an isomorphism $\varphi : T_1 \to T_2$.

Conversely, suppose that $\varphi : T_1 \to T_2$ is an isomorphism. Then we have $\varphi \pi^1 : S_1 \to T_2$, where $\pi^1 : S_1 \to T_1$ is the projection, and we have the projection $\pi^2 : S_2 \to T_2$, which is a surjection. Since S_1 is projective, we have a map $f : S_1 \to S_2$ with $\pi^2 f = \varphi \pi^1$. Note $\mathrm{Im}(f)$ is not contained in J_2, and since J_2 is the unique maximal proper left R-ideal in S_2, we have that f is surjective. Similarly, from $\varphi^{-1} : T_2 \to T_1$ we may obtain a surjective map $g : S_2 \to S_2$. Then the composition $gf : S_1 \to S_1$ is surjective, and hence it is an isomorphism by Lemma 1.11, and so f is an isomorphism.

(3) Clearly each T_i is simple.

Let M be an irreducible R-module. Then $M = RM \neq \{0\}$, so $S_i M \neq \{0\}$ for some i, and so $e_i m \neq 0$ for some m. Let $\theta : S_i \to M$ by $\theta(r) = rm$. Since M is simple, $\theta(S_i) = M$, so M is isomorphic to $S_i / \mathrm{Ker}(\theta)$. Again, since M is simple, $\mathrm{Ker}(\theta)$ must be a maximal proper R-ideal in S_i, so $\mathrm{Ker}(\theta) = J_i$ and M is isomorphic to $S_i / J_i = T_i$. $\qquad\square$

We use the notation S_i, J_i, and T_i henceforth. We let $D_i = \mathrm{Hom}_R(T_i, T_i)$ and note that D_i is a division algebra by Schur's lemma.

Notation. Let t' denote the number of isomorphism classes of principal indecomposable, or simple, R-modules.

Corollary 3.10. *Let M be a direct sum of principal indecomposable R-modules. Then M is semisimple if and only if $J(R)M - \{0\}$.*

Proof. We have already seen in Proposition 2.9 that M simple implies $J(R)M = \{0\}$, and this immediately extends to M semisimple. Conversely, write $M \cong \bigoplus M_j$, where M_j is indecomposable. Then $J(R)M = \{0\}$ implies $J(R)M_j = \{0\}$ for each j. Now M_j is isomorphic to S_i for some i, so $\{0\} = J(R)S_i = J_i$, so $S_i \cong S_i / J_i = T_i$ is simple, and M is a direct sum of simple modules, so by Corollary 2.2.6 M is semisimple. $\qquad\square$

Corollary 3.11. *Let R be a ring. Then R is semisimple if and only if R is J-semisimple and of finite length.*

Proof. We have already seen that R semisimple implies R J-semisimple (Proposition 2.10) and of finite length (Corollary 2.2.15). Conversely, let R be J-semisimple and of finite length. Then apply Theorem 1.19 (1) and then Corollary 3.10 with $M = R$ to conclude that R is semisimple. $\qquad\square$

Corollary 3.12. *Let R be a ring of finite length, and let M be an R-module. Then M is semisimple if and only if $J(R)M = \{0\}$.*

Proof. If M is semisimple, then $J(R)M = \{0\}$ by Proposition 2.9.

Conversely, if $J(R)M = \{0\}$, then M is an $R/J(R)$-module. But the ring $R/J(R)$ is J-semisimple and has finite length, as R does, so is semisimple. Thus by Theorem 2.2.12 M is semisimple as an $R/J(R)$-module, and hence as an R-module as well. $\qquad\square$

Lemma 3.13.

(1) *Let M be any simple R-module. Then any $f \in \mathrm{Hom}_R(S_i, M)$ factors through $\pi : S_i \to T_i$.*

(2) *If M is a simple R-module, then $\mathrm{Hom}_R(S_i, M) = \{0\}$ if M is not isomorphic to T_i and $\mathrm{Hom}_R(S_i, M) \cong D_i$ if M is isomorphic to T_i.*

Proof.

(1) Since M is simple, and J_i is the unique maximal ideal of S_i, $\mathrm{Ker}(f)$ contains J_i.

(2) This is immediate from (1) and Schur's lemma. $\qquad\square$

Proposition 3.14. *Let M be an R-module of finite length. Then*

$$\mathrm{Hom}_R(S_i, M) \cong D_i^c,$$

where c is the number of composition factors of M that are isomorphic to T_i.

Proof. Let M have composition series

$$\{0\} \subset \cdots \subset M_{\ell-1} \subset M_\ell = M.$$

Let $H_k = \{f \in \mathrm{Hom}_R(S_i, M) \mid \mathrm{Im}(f) \subset M_k\}$. Then

$$(3.1) \qquad\qquad \{0\} \subseteq \cdots \subseteq H_{\ell-1} \subseteq H_\ell = \mathrm{Hom}_R(S_i, M).$$

If M_k/M_{k-1} is not isomorphic to T_i, then $\mathrm{Hom}_R(S_i, M_k/M_{k-1}) = \{0\}$ by Lemma 3.13, which implies that $H_k/H_{k-1} = \{0\}$.

If M_k/M_{k-1} is isomorphic to T_i, then $\mathrm{Hom}_R(S_i, M_k/M_{k-1}) \cong D_i$, also by Lemma 3.13. But if $\overline{f} \in \mathrm{Hom}_R(S_i, M_k/M_{k-1})$, then $\overline{f} = \pi f$ for some $f \in \mathrm{Hom}_R(S_i, M_k)$, as $\pi : M_k \to M_k/M_{k-1}$ is surjective and S_i is a projective R-module. Thus in this case $H_k/H_{k-1} \cong D_i$.

Hence (3.1) has c nonzero terms, each of which is isomorphic to D_i, and the proposition follows. $\qquad\square$

Now we turn to the study of block decompositions of R.

Recall that we introduced the notion of an acceptable ring in Definition 2.3.9.

Lemma 3.15. *Let R be a ring of finite length. Then R is acceptable.*

Proof. Immediate from Theorem 1.19 (1). □

In light of this lemma we may apply the results of Section 2.3. In particular, we may immediately conclude:

Lemma 3.16. *Let R be a ring of finite length. Then R has a unique block decomposition.*

Proof. Immediate from Lemma 3.15 and Theorem 2.3.15. □

We wish to obtain more precise information on block decompositions.

Corresponding to the decomposition $R = \bigoplus S_i$ we have the complementary set of primitive idempotents $\{e_i\}$ with $S_i = Re_i$ (Corollary 2.3.7). Further, R has a unique complementary set of central-primitive idempotents $\{e_\alpha\}$, with $R = \bigoplus Re_\alpha = \bigoplus B_\alpha$ the block decomposition of R (Lemma 2.3.14 and Theorem 2.3.15).

Recall from Lemma 2.3.11 that $e_\alpha = \sum a_{\alpha i} e_i$ with each $a_{\alpha i} = 0$ or 1. Then, as $\{e_\alpha\}$ is complementary, for each fixed i there is a unique α with $a_{\alpha i} = 1$. In this case we say i belongs to α, and this is equivalent to S_i being contained in B_α.

In the semisimple case, S_i and S_j were contained in the same block if and only if they were isomorphic. In general, however, the condition for being contained in the same block is weaker.

Definition 3.17. Two principal indecomposable R-modules S_i and S_j are **linked** if there is a sequence of principal indecomposable R-modules $S_{i_1}, \ldots, S_{i_m}$ with $S_i = S_{i_1}$, $S_j = S_{i_m}$, and $\mathrm{Hom}_R(S_{i_k}, S_{i_{k+1}}) \neq \{0\}$ or $\mathrm{Hom}_R(S_{i_{k+1}}, S_{i_k}) \neq \{0\}$ for each $k = 1, \ldots, m-1$.

Proposition 3.18. *Two principal indecomposable R-modules S_i and S_j are contained in the same block B_α, for some α, if and only if S_i and S_j are linked.*

Proof. Observe that S_i is contained in B_α if and only if $e_\alpha e_i \neq 0$. Suppose that S_i is contained in B_α and that $\mathrm{Hom}_R(S_i, S_j) \neq \{0\}$. Let $\varphi : S_i \to S_j$ be nonzero. Then $0 \neq \varphi(e_i) = r_0 e_j$ for some r_0. Then $0 \neq \varphi(e_i) = \varphi(e_\alpha e_i) = e_\alpha \varphi(e_i) = e_\alpha(r_0 e_j) = r_0(e_\alpha e_j)$ so $e_\alpha e_j \neq 0$ and S_j is also contained in B_α. Hence if S_i and S_j are linked, they are contained in the same block.

Conversely, suppose B_α is a block, and suppose not all S_i in B_α are linked. We shall show that e_α is not central primitive, a contradiction. Let $\{S_i'\}$ be an equivalence class of $\{S_i$ contained in $B_\alpha\}$ under the equivalence relation of linkage, with $\{S_i''\}$ its complement in $\{S_i$ contained in $B_\alpha\}$. Let $\{e_i'\}$ and $\{e_i''\}$ be the corresponding sets of idempotents, and set $f_1 = \sum e_i'$, $f_2 = \sum e_i''$. Then $e_\alpha = f_1 + f_2$. Observe that $\varphi : Re_i' \to Re_i''$ by $\varphi(r) = re_i''$ is a homomorphism, and since no S_i' and S_i'' are linked, it must be zero, so in particular $0 = \varphi(e_i') = e_i'e_i''$. Hence $f_1 f_2 = f_2 f_1 = 0$, so f_1 and f_2 are orthogonal. We claim that f_1 and f_2 are central. Since $Rf_1 \subseteq Re_\alpha$ and $Rf_2 \subseteq Re_\alpha$, and $\{e_\alpha\}$ are orthogonal, it suffices to show that f_1 and f_2 are central in Re_α. Now, for any $r_0 \in R$, $\varphi : Rf_1 \to Rf_2$ by $\varphi(r) = rr_0 f_2$ is also the zero map, so in particular $0 = \varphi(f_1) = f_1 r_0 f_2$. Let $r \in Re_\alpha$ and write $r = r_0 e_\alpha$. Then $f_1 r = f_1 r_0 e_\alpha = f_1 r_0 (f_1 + f_2) = f_1 r_0 f_1 + f_1 r_0 f_2 = f_1 r_0 f_1$, and similarly $rf_1 = r_0 e_\alpha f_1 = e_\alpha r_0 f_1 = f_1 r_0 f_1$, so $f_1 r = f_1 r_0 f_1 = rf_1$ and f_1 is central, and similarly for f_2, completing the proof. $\square$

Definition 3.19. Two irreducible R-modules T_i and T_j are **in the same block** if the corresponding principal indecomposable R-modules S_i and S_j are contained in the same block.

We will not need the rest of this section, but for the sake of general mathematical knowledge we will present it.

Definition 3.20.

(1) An R-module M is **artinian**, or satisfies the **descending chain condition**, if every chain of submodules

$$M = M_0 \supset M_1 \supset M_2 \supset \cdots$$

eventually terminates (necessarily with $M_k = \{0\}$ for some k).

(2) An R-module M is **noetherian**, or satisfies the **ascending chain condition**, if every chain of submodules

$$\{0\} = M_0 \subset M_1 \subset M_2 \subset M_3 \subset \cdots$$

eventually terminates (necessarily with $M_k = M$ for some k).

(3) *A ring R is* **artinian** *if R is artinian as an R-module, and is* **noetherian** *if R is noetherian as an R-module.*

Proposition 3.21. *An R-module M is of finite length if and only if M is both artinian and noetherian.*

Proof. $\square$

We are interested in algebras of finite dimension over a field, and it is clear that they have finite length. Thus we may as well make that hypothesis. However, although it was not known to E. Artin and E. Noether, it turns out that every artinian ring is noetherian, and hence of finite length. (The converse is not true. For example, $\mathbb{Z}$ is noetherian but not artinian.) Because of this fact, theorems in this subject are usually stated with the hypothesis "*Let R be an artinian ring.*" We now proceed to prove this fact.

Lemma 3.22. *Let R be artinian. Then $J(R)$ is nilpotent.*

Proof. This is Proposition 3.2. Although we stated it with the hypothesis that R had finite length, the proof only used the descending chain condition. $\square$

Lemma 3.23. *Let R be J-semisimple and artinian. Then R has finite length. Indeed, R is semisimple.*

Proof. We begin with two observations. Let I be a nonzero left ideal of R. Then:

(1) I contains a minimal nonzero left ideal. This is just the descending chain condition.

(2) If $I_{\min}$ is a minimal nonzero left ideal of I, then $I_{\min}$ is a summand of I. For $J(R) = \{0\}$, so there is a maximal ideal $I_{\max}$ of R with $I_{\min} \not\subseteq I_{\max}$, and then $I = I_{\min} \oplus (I_{\max} \cap I)$. (Note $I_{\min} \cap I_{\max} = \{0\}$ as $I_{\min}$ is minimal, and $I_{\min} + I_{\max} = R$ as $I_{\max}$ is maximal. If $i \in I$, write $i = i_{\min} + i_{\max}$, so in this case $i_{\max} = i - i_{\min} \in I$.)

Let $I_0 = R$. Then $I_0 = I_1 \oplus I_1'$ by (1) and (2) where I_1 is minimal (and hence simple). Then, similarly, $I_1' = I_2 \oplus I_2'$ with I_2 minimal (and hence simple). Proceeding in this way we get a chain of ideals $R \supset I_1' \supset I_2' \supset \cdots$, and by the descending chain condition this must stop, so $I_k' = \{0\}$ for some k, and then $R = I_1 \oplus I_2 \oplus \cdots \oplus I_k$ has length k. Also, R is a sum of simple modules so, by Corollary 2.2.6, R is semisimple. $\square$

Theorem 3.24 (Hopkins-Levitzki). *Let R be a ring such that $J(R)$ is nilpotent and $R/J(R)$ is semisimple. Let M be an R-module. Then M is artinian if and only if M is noetherian if and only if M has finite length.*

Proof. Only one direction needs proof: if M satisfies either chain condition it is of finite length.

Let $J = J(R)$. Then J is nilpotent by hypothesis, so $J^\ell = \{0\}$ for some ℓ. Thus there is a finite chain

$$\{0\} = J^\ell M \subset \cdots \subset J^2 M \subset JM \subset M.$$

Now each $J^k M / J^{k+1} M$ is an R/J-module that satisfies the same chain condition M does. But R/J is semisimple so $J^k M / J^{k+1} M$ is a direct sum of simple R/J modules, and either chain condition implies that this sum is finite. Hence M has finite length.   □

Corollary 3.25. *Let R be an artinian ring. Then R is noetherian. Also, R has finite length.*

Proof. Since R is artinian, $J(R)$ is nilpotent by Lemma 3.22. Since R is artinian, $R/J(R)$ is artinian (as any chain of $R/J(R)$-modules may be regarded as a chain of R-modules). Thus $R/J(R)$ is J-semisimple and artinian, hence it is semisimple by Lemma 3.23.

Thus R satisfies the hypotheses of Theorem 3.24, so that theorem applied to the case $M = R$ shows that R is noetherian, or of finite length.   □

Remark. Our exposition here follows Lam [La, Theorem 4.15].

We record the following generalization of Lemma 3.15.

Lemma 3.26. *If R is artinian or noetherian, then R is acceptable.*

Proof. We must show that R is a finite direct sum of indecomposable modules. Assume not. Then R is decomposable, $R = I_1 \oplus J_1$, with $I_1 \neq \{0\}$, $J_1 \neq \{0\}$, where at least one of I_1 and J_1 is not a finite direct sum of indecomposable modules. Call that one J_1. Then, similarly, $J_1 = I_2 \oplus J_2$ where J_2 is not a finite sum of indecomposables. Continuing in this way, we get infinite sequences of submodules $\{I_k\}$ and $\{J_k\}$. Then

$$\cdots \subset J_3 \subset J_2 \subset J_1 \subset R$$

is an infinite descending chain, contradicting the descending chain condition. Similarly,

$$\{0\} \subset I_1 \subset I_1 \oplus I_2 \subset I_1 \oplus I_2 \oplus I_3 \subset \cdots$$

is an infinite ascending chain, contradicting the ascending chain condition.□

Remark. Of course, we have shown that artinian implies noetherian, but we have chosen to present a direct proof of both halves of this lemma anyway.

Remark 3.27. While artinian implies noetherian for rings, these are independent properties for modules: As a $\mathbb{Z}$-module, $\mathbb{Z}$ is noetherian but not artinian, since it has the infinite descending chain

$$\cdots \subset 8\mathbb{Z} \subset 4\mathbb{Z} \subset 2\mathbb{Z} \subset \mathbb{Z},$$

and as a $\mathbb{Z}$-module, $\mathbf{Q}/\mathbb{Z}$ is artinian but not noetherian, since it has the infinite ascending chain

$$\{0\} \subset (\tfrac{1}{2}\mathbb{Z})/\mathbb{Z} \subset (\tfrac{1}{4}\mathbb{Z})/\mathbb{Z} \subset (\tfrac{1}{8}\mathbb{Z})/\mathbb{Z} \subset \cdots .$$

6.4. Finite-dimensional Algebras

Now we specialize the case of rings of finite length further, to the case of finite-dimensional algebras over a field.

We begin by showing that if R is a finite-dimensional algebra over the field $\mathbf{F}_p$, then the notions of splitting field and universal field of definition for R coincide, and that $\overline{\mathbf{F}}_p$, or even some finite extension of $\mathbf{F}_p$, is a splitting field for R. This parallels the semisimple case. Letting $\overline{R}$ denote $\overline{\mathbf{F}}_p \otimes_{\mathbf{F}_p} R$, we find a formula for the number t' of isomorphism classes of simple $\overline{R}$-modules, and a formula for $\sum d_i^2$, where $\{d_i\}$ are the dimensions of these modules, also paralleling the semisimple case. Finally, we generalize theorems of Burnside and Frobenius-Schur to this case.

Proposition 4.1.

(1) *Let R be an $\mathbf{F}_p$-algebra, let $\mathbf{F}$ be an arbitrary field of characteristic p, and let $R' = \mathbf{F} \otimes_{\mathbf{F}_p} R$. Then $J(R') \supseteq \mathbf{F} \otimes_{\mathbf{F}_p} J(R)$.*

(2) *If R is a finite-dimensional $\mathbf{F}_p$-algebra, then $J(R') = \mathbf{F} \otimes_{\mathbf{F}_p} J(R)$, and hence $R'/J(R') \cong \mathbf{F} \otimes_{\mathbf{F}_p} (R/J(R))$.*

Proof.

(1) $J(R)$ is a nilpotent ideal in R, so $\mathbf{F} \otimes_{\mathbf{F}_p} J(R)$ is a nilpotent ideal in R', so $\mathbf{F} \otimes_{\mathbf{F}_p} J(R) \subseteq J(R')$.

(2) As $R/J(R)$ is semisimple in this case, we have that

$$R/J(R) \cong \bigoplus \mathrm{Mat}_{n_i}(D_i)$$

by Wedderburn's theorem (Theorem 2.2.16), where D_i is a finite division ring, and hence, by another theorem of Wedderburn (Theorem A.1.1), a field $\mathbf{F}_i$.

Now for any two-sided ideal I in R, $\mathbf{F} \otimes_{\mathbf{F}_p} (R/I) \cong (\mathbf{F} \otimes_{\mathbf{F}_p} R)/(\mathbf{F} \otimes_{\mathbf{F}_p} I)$; in particular this is true for $I = J(R)$. Hence

$$(\mathbf{F} \otimes_{\mathbf{F}_p} R)/(\mathbf{F} \otimes_{\mathbf{F}_p} J(R)) \cong \mathbf{F} \otimes_{\mathbf{F}_p} (R/J(R))$$

$$\cong \bigoplus \mathbf{F} \otimes_{\mathbf{F}_p} \mathrm{Mat}_{n_i}(\mathbf{F}_i)$$

$$\cong \bigoplus \mathrm{Mat}_{n_i}(\mathbf{F} \otimes_{\mathbf{F}_p} \mathbf{F}_i).$$

But, by Theorem A.1.5, $\mathbf{F} \otimes_{\mathbf{F}_p} \mathbf{F}_i \cong \bigoplus \mathbf{F}_{ij}$, a direct sum of fields, and so $\mathbf{F} \otimes_{\mathbf{F}_p} (R/J(R))$ is isomorphic to a direct sum of matrix rings over fields, and hence is semisimple, so in particular is J-semisimple.

Thus $\mathbf{F} \otimes_{\mathbf{F}_p} J(R) \supseteq J(R')$, and so they are equal. $\qquad\square$

Corollary 4.2. *Let R be a finite-dimensional $\mathbf{F}_p$-algebra, let $\mathbf{F}$ be an arbitrary field of characteristic p, and let $R' = \mathbf{F} \otimes_{\mathbf{F}_p} R$. Then*

(1) *The $\mathbf{F}$-algebra $\mathbf{F} \otimes_{\mathbf{F}_p} (R/J(R))$ is semisimple.*

(2) *Let M be an R-module and let $M' = \mathbf{F} \otimes_{\mathbf{F}_p} M$. Then M is semisimple if and only if M' is semisimple.*

Proof.

(1) Since $R/J(R)$ is J-semisimple, this follows from Proposition 4.1 (2).

(2) M is semisimple if and only if $J(R)M = \{0\}$ and M' is semisimple if and only if $J(R')M' = (\mathbf{F} \otimes_{\mathbf{F}_p} J(R))(\mathbf{F} \otimes_{\mathbf{F}_p} M) = \{0\}$. Clearly these two conditions are equivalent. $\qquad\qquad\square$

Remark 4.3.

(1) Proposition 4.1 generalizes, with the same proof, if R is an algebra over any finite field of characteristic p, and $\mathbf{F}$ is any field of characteristic p containing that field.

(2) Part (2) of Proposition 4.1 is in general false without the assumption that R is finite dimensional. Here is a counterexample: Let $\mathbf{F}_p(x)$ denote the field of rational functions in the variable x over the field $\mathbf{F}_p$. Let $R = \mathbf{F} = \mathbf{F}_p(x)[t]/(t^p - x)$. Then R is an inseparable extension of $\mathbf{F}_p[x]$ of degree p, and $J(R) = \{0\}$ as R is a field. Then

$$\begin{aligned}
R' &= \mathbf{F} \otimes_{\mathbf{F}_p} R \\
&= \mathbf{F}_p(x)[t]/(t^p - x) \otimes_{\mathbf{F}_p(x)} \mathbf{F}_p(x)[u]/(u^p - x) \\
&= (\mathbf{F}_p(x)[t]/(t^p - x))[u]/(u^p - x) \\
&= (\mathbf{F}_p(x)[t]/(t^p - x))[u]/[u^p - t^p] \\
&= (\mathbf{F}_p(x)[t]/(t^p - x))[u]/[(u - t)^p]
\end{aligned}$$

which contains the nilpotent ideal $(u - t)$, so $J(R') \neq \{0\}$.

Definition 4.4. Let R be a finite-dimensional $\mathbf{F}_p$-algebra.

(1) $\mathbf{F}$ is a **splitting field** for R if every irreducible $\mathbf{F} \otimes_{\mathbf{F}_p} R$-module is absolutely irreducible.

(2) $\mathbf{F}$ is a **universal field of definition** for R if for any field $\mathbf{F}'$ containing $\mathbf{F}$, every irreducible $\mathbf{F}' \otimes_{\mathbf{F}_p} R$-module is defined over $\mathbf{F}$.

Note that part (1) of this definition agrees with that in Section 2.4, while in part (2) we restrict ourselves to irreducible modules, in light of Example 5.1.3, which shows that no such $\mathbf{F}$ will in general exist if we allow all modules.

Theorem 4.5. *Let R be a finite-dimensional $\mathbf{F}_p$-algebra.*

(1) $\mathbf{F}$ *is a splitting field for R if and only if $\mathbf{F}$ is a universal field of definition for R.*

(2) $\overline{\mathbf{F}}_p$ *is a splitting field for R.*

(3) *There is some finite field $\mathbf{F}$ that is a splitting field for R.*

Proof. By Corollary 3.12, simple R-modules are in one-to-one correspondence with simple $R/J(R)$-modules. But now, by Corollary 4.2, we may apply the results of Section 2.4 to $R/J(R)$, so the theorem follows by results 2.4.15, 2.4.12, and 2.4.16. $\qquad\square$

Proposition 4.6. *Let R be a finite-dimensional algebra over an algebraically closed field $\mathbf{F}$. Then there are only finitely many simple R-modules, up to isomorphism. If their degrees are $\{d_i\}$, then*

$$\sum d_i^2 = \dim_{\mathbf{F}} R - \dim_{\mathbf{F}} J(R).$$

Proof. In this case $R/J(R)$ is semisimple, so this is immediate from Theorem 4.5 and Corollary 2.2.24. $\qquad\square$

Remark. Since $J(R)$ is difficult to compute, this is not a practical method for computing $\sum d_i^2$. Indeed, this proposition is most useful in the other direction, to aid in determining $J(R)$ once all the simple R-modules are known.

Now we determine the number t' of distinct simple R-modules.

Lemma 4.7. *Let R be an $\mathbf{F}$-algebra, $\mathrm{char}(\mathbf{F}) = p$. Let $L \subseteq R$ be the $\mathbf{F}$-vector space spanned by $\{r_1 r_2 - r_2 r_1 \mid r_1, r_2 \in R\}$. Then*

(1) *For any $r_1, \dots, r_m \in R$ and any $i \geq 0$,*

$$(r_1 + \cdots + r_m)^{p^i} \equiv r_1^{p^i} + \cdots + r_m^{p^i} \pmod{L}.$$

(2) *If $r \in L$, then $r^{p^i} \in L$ for every $i \geq 0$.*

Proof.

(1) If R were commutative, then the result would be immediate from the multinomial theorem. In the general case we must proceed more carefully. Arguing by induction, it suffices to prove this in case $i = 1$. Regard $r_1, \dots, r_m$ as noncommuting indeterminates. Then $(r_1 + \cdots + r_m)^p$ consists of m^p distinct words in $r_1, \dots, r_m$, each of length p. Let the cyclic group of order p act on this set of words by "rotation". The words $r_1^p, \dots, r_m^p$ are fixed under this action, while the others are permuted freely. In particular,

any other orbit has cardinality p. But any two words in the same orbit are congruent modulo L (for example, $abcde - eabcd = (abcd)e - e(abcd) \in L$), so the sum of the words in such an orbit is congruent to $pR \pmod{L}$, and $pR = \{0\}$.

(2) Again it suffices to prove this in case $i = 1$. Let $r = r_1 r_2 - r_2 r_1$. Then, by part (1),

$$
\begin{aligned}
r^p &\equiv (r_1 r_2)^p - (r_2 r_1)^p \pmod{L} \\
&= r_1\big((r_2 r_1)^{p-1} r_2\big) - \big((r_2 r_1)^{p-1} r_2\big) r_1 \pmod{L} \\
&\equiv 0 \pmod{L},
\end{aligned}
$$

so $r^p \in L$. If r is a sum of elements of this form, then another application of (1) shows $r^p \in L$ in this case, too. $\qquad\square$

Remark. Note that L is in general *not* an R-ideal.

Lemma 4.8. *Let R be an $\mathbf{F}$-algebra, $\mathrm{char}(\mathbf{F}) = p$. Let $M \subseteq R$ be*

$$
M = \{r \in R \mid r^{p^i} \in L \text{ for some } i \geq 0\}.
$$

Then M is an $\mathbf{F}$-vector space.

Furthermore,

(1) *If $R = \mathrm{Mat}_d(\mathbf{F})$, then*

$$
M = L = \{r \in R \mid \mathrm{Tr}(r) = 0\},
$$

and $\dim_{\mathbf{F}} R - \dim_{\mathbf{F}} L = 1$.

(2) *If $\mathbf{F}$ is a splitting field for R and R is semisimple, then $M = L$.*

(3) *If $\mathbf{F}$ is a splitting field for R, then $M = J(R) + L$, where $J(R) + L$ is the $\mathbf{F}$-vector space (not the R-ideal) generated by $J(R)$ and L.*

Proof. Suppose $r_1, r_2 \in M$. Choose i so that $r_1^{p^i} \in L$, $r_2^{p^i} \in L$. Then $(r_1 + r_2)^{p^i} \in L$ by Lemma 4.7, so M is an $\mathbf{F}$-vector space. Observe that $M \supseteq L$ by definition.

(1) Let $R = \mathrm{Mat}_d(\mathbf{F})$. As $\mathrm{Tr}(r_1 r_2) = \mathrm{Tr}(r_2 r_1)$, clearly $L \subseteq \{r \in R \mid \mathrm{Tr}(r) = 0\}$. On the other hand,

$$
\begin{bmatrix} 0 & 1 \\ 0 & 0 \end{bmatrix} = \begin{bmatrix} 1 & 0 \\ 0 & 0 \end{bmatrix}\begin{bmatrix} 0 & 1 \\ 0 & 0 \end{bmatrix} - \begin{bmatrix} 0 & 1 \\ 0 & 0 \end{bmatrix}\begin{bmatrix} 1 & 0 \\ 0 & 0 \end{bmatrix},
$$

$$
\begin{bmatrix} 0 & 0 \\ 1 & 0 \end{bmatrix} = \begin{bmatrix} 0 & 0 \\ 1 & 0 \end{bmatrix}\begin{bmatrix} 1 & 0 \\ 0 & 0 \end{bmatrix} - \begin{bmatrix} 1 & 0 \\ 0 & 0 \end{bmatrix}\begin{bmatrix} 0 & 0 \\ 1 & 0 \end{bmatrix},
$$

$$\begin{bmatrix} 1 & 0 \\ 0 & -1 \end{bmatrix} = \begin{bmatrix} 0 & 1 \\ 0 & 0 \end{bmatrix}\begin{bmatrix} 0 & 0 \\ 1 & 0 \end{bmatrix} - \begin{bmatrix} 0 & 0 \\ 1 & 0 \end{bmatrix}\begin{bmatrix} 0 & 1 \\ 0 & 0 \end{bmatrix},$$

proving the reverse inclusion in case $d = 2$, and the general case is similar. It remains to prove $M = L$. Now $\dim_{\mathbf{F}} R = d^2$, $\dim_{\mathbf{F}}$ {matrices of trace 0} $= d^2 - 1$, so we need only show $M \neq R$. But this is clear as $\begin{bmatrix} 1 & 0 \\ 0 & 0 \end{bmatrix} \in R$, $\begin{bmatrix} 1 & 0 \\ 0 & 0 \end{bmatrix} \notin M$ in case $d = 2$, and similarly for d in general.

(2) Now let R be semisimple. Then $R \cong \bigoplus \mathrm{Mat}_{d_i}(D_i)$ by Wedderburn's theorem, and if $\mathbf{F}$ is a splitting field for R, then each $D_i = \mathbf{F}$, so this case follows from (1).

(3) Note that $J(R) \subseteq M$, as $J(R)$ is nilpotent. Let $\overline{R} = R/J(R)$, and note that $\overline{L} = \{\overline{r}_1 \overline{r}_2 - \overline{r}_1 \overline{r}_1 \mid \overline{r}_1, \overline{r}_2 \in \overline{R}\}$ is the image of L under the projection $R \to \overline{R}$. Thus $M/J(R) = \overline{L}$ by (2), as $\overline{R}$ is semisimple, so (3) follows. $\qquad\square$

Remark. Note that M is in general *not* an R-ideal.

Theorem 4.9. *Let $\mathbf{F}$ be a splitting field for R, and let $\dim_{\mathbf{F}} R$ be finite. Then $t'(R)$, the number of distinct isomorphism classes of simple R-modules, is given by*

$$t'(R) = \dim_{\mathbf{F}} R - \dim_{\mathbf{F}} M,$$

where M is as in Lemma 4.8.

Proof. Let $\overline{R} = R/J(R)$. Then there is a one-to-one correspondence between isomorphism classes of simple R-modules and of simple $\overline{R}$-modules, by Corollary 3.12. Now under the projection $R \to \overline{R}$, the image of M is $\overline{L}$, by Lemma 4.8 (3), (and in the notation of the proof there). Thus $\dim_{\mathbf{F}} R - \dim_{\mathbf{F}} M = \dim_{\mathbf{F}} \overline{R} - \dim_{\mathbf{F}} \overline{L}$. Now $\overline{R}$ is semisimple. Hence $\overline{R} = \bigoplus_{i=1}^{t'(R)} \mathrm{Mat}_{d_i}(\mathbf{F})$, as in the proof of Lemma 4.8 (2). But then by Lemma 4.8 (1), $\dim_{\mathbf{F}} \overline{R} - \dim_{\mathbf{F}} \overline{L} = t'(R) \cdot 1 = t'(R)$. $\qquad\square$

Theorem 4.10. *Let R be a finite-dimensional $\mathbf{F}$-algebra, where $\mathbf{F}$ is an algebraically closed field. Then the theorems of Burnside (Theorem 2.5.3) and Frobenius-Schur (Theorem 2.5.6) hold for R.*

Proof. We proved these in Section 2.5 under the hypothesis that R was semisimple. Replacing R by $R/J(R)$, which is semisimple, and recalling that every semisimple R-module is an $R/J(R)$-module, the proofs there go through unchanged. $\qquad\square$

Chapter 7

Modular Group Representations

7.1. General Results

First recall our notational conventions: G is a finite group of order n divisible by the prime p, and we write $n = p^r q$ with p not dividing q. P is a p-Sylow subgroup of G, and $\mathbf{F}$ is a field of characteristic p.

We begin by reminding the reader what we know about p-groups.

Proposition 1.1. *Let G be a p-group.*

(1) *The only simple $\mathbf{F}(G)$-module is $\mathbf{F}$, with trivial G-action (up to isomorphism).*

(2) *$\mathbf{F}(G)$ is an indecomposable $\mathbf{F}(G)$-module.*

Proof. This is just Example 5.1.2, which we have restated here for convenience (and emphasis). $\qquad\square$

Thus in this situation we see that we have one irreducible $\mathbf{F}(G)$-module, namely τ, and one principal indecomposable $\mathbf{F}(G)$-module, namely $\mathbf{F}(G)$ itself. Note τ appears p^r times in the composition series for $\mathbf{F}(G)$.

Now we turn to more general groups.

Lemma 1.2. *Let V be a simple $\mathbf{F}(G)$-module and let H be any normal p-subgroup of G. Then H acts trivially on V.*

Proof. Let $W = \operatorname{Res}^G_H(V)$. By Corollary 4.1.12, W is a semisimple $\mathbf{F}(H)$-module, so it is a direct sum of simple $\mathbf{F}(H)$-modules. But the only simple $\mathbf{F}(H)$-module is $\mathbf{F}$, with the trivial action of H, by Proposition 1.1. □

Corollary 1.3. *Suppose that P is a normal subgroup of G, and let $\pi : G \to G/P$ be the quotient map. Then π^* (pullback under π) induces a one-to-one correspondence between simple $\mathbf{F}(G/P)$-modules and simple $\mathbf{F}(G)$-modules.*

Proof. □

Corollary 1.4. *Suppose that P is a normal subgroup of G. Then the semisimple $\mathbf{F}$-representations of G are in one-to-one correspondence with the $\mathbf{F}$-representations of G/P. In particular, if $\mathbf{F}$ is algebraically closed, the number t' of simple $\mathbf{F}$-representations of G is equal to the number of conjugacy classes in G/P, their degrees $d_1, \dots, d_{t'}$ satisfy $\sum d_i^2 = q$, and d_i divides q for each i.*

Proof. This is just a special case of Corollary 1.3 combined with the observation that the order q of G/P is prime to p, so that $\mathbf{F}(G/P)$ is semisimple. □

We emphasize that this corollary shows that in case P is normal, the theory of semisimple $\mathbf{F}(G)$ modules reduces to the theory of $\mathbf{F}(G/P)$-modules, · and $\mathbf{F}(G/P)$ is a semisimple ring, so we are in a familiar situation.

We pause here to record some information about the Jacobson radical of $\mathbf{F}(G)$.

Corollary 1.5.

(1) **(Wallace)** *Let $R = \mathbf{F}(G)$ where G is a p-group. Then $J(R) = \mathcal{R}_0$, the augmentation ideal of R, and $\dim_{\mathbf{F}} J(R) = p^r - 1$.*

(2) *Suppose that P is a normal subgroup of G and let $R = \mathbf{F}(G)$. Then $J(R)$ is the ideal generated by $J(\mathbf{F}(P))$ (under the natural inclusion of $\mathbf{F}(P)$ in $\mathbf{F}(G)$), and $\dim_{\mathbf{F}} J(R) = q(p^r - 1)$.*

(3) *Let H be any normal p-subgroup of G. Then $J(R)$ contains the ideal generated by $J(\mathbf{F}(H))$. In particular, this holds when H is the intersection of all the p-Sylow subgroups of G.*

Proof. We first prove this under the assumption that $\mathbf{F}$ is algebraically closed. The general case then follows from Proposition 6.4.1.

(1) Clearly $\dim_{\mathbf{F}} \mathcal{R}_0 = p^r - 1$. Note that $\mathcal{R}_0$ is a maximal left ideal, so $J(R) \subseteq \mathcal{R}_0$. On the other hand, by Proposition 6.4.6, $\dim_{\mathbf{F}} R - \dim_{\mathbf{F}} J(R) = \sum d_i^2$, where $\{d_i\}$ are the degrees of the irreducible $\mathbf{F}$-representations of G. But there is only one of these, the trivial representation τ of degree 1, by Proposition 1.1. Hence, $J(R) = \mathcal{R}_0$.

(2) Observe that $\mathcal{R}_0$ is generated by $\{g - 1 \mid g \in G\}$. Thus part (1), applied to the case $G = P$, shows that $J(\mathbf{F}(P))$ is generated by $\{a - 1 \mid a \in P\}$, and it is easy to check from this that $\dim_{\mathbf{F}} RJ(\mathbf{F}(P)) = q(p^r - 1)$. Now for any $g_1, g_2 \in G$, $a_1, a_2 \in P$, $g_1(a_1 - 1)g_2(a_2 - 1) = g_1 g_2(a'_1 - 1)(a_2 - 1)$, where $a'_1 = g_2^{-1} a_1 g_2 \in P$, as P is normal. Furthermore, we know by Proposition 6.3.2 that $J(\mathbf{F}(P))$ is nilpotent, so we see that $RJ(\mathbf{F}(P))$ is nilpotent as well. Hence $RJ(\mathbf{F}(P)) \subseteq J(R)$.

We again apply Proposition 6.4.6. $\dim_{\mathbf{F}} R - \dim_{\mathbf{F}} J(R) = \sum d_i^2$, where $\{d_i\}$ are the irreducible $\mathbf{F}$-representations of G. But these correspond to the $\mathbf{F}$-representations of G/P by Corollary 1.4, so $\sum d_i^2 = q$, and so $\dim_{\mathbf{F}} J(R) = q(p^r - 1)$. Hence, $RJ(\mathbf{F}(P)) = J(R)$.

(3) The first paragraph of the proof of part (2) applied to H yields part (3).
$\square$

Lemma 4.5.5 and Definition 4.5.6 are key tools in modular representation theory. For the convenience of the reader we repeat them here.

Lemma 1.6. *Fix a prime number p. Then any $g \in G$ can be written uniquely as $g = ab$, where the order of a is prime to p, the order of b is a power of p, and a and b commute. In this situation both a and b are powers of g.*

Proof. If $(n, p) = 1$, the proof is trivial. If $n = p^r q$ with $r \geq 1$ and $(p, q) = 1$, then set $1 = xp^r + yq$ for some integers x and y. Set $a = g^{xp^r}$ and $b = g^{yq}$. Uniqueness is easy to check. $\square$

Definition 1.7. In the situation of Lemma 1.6, a is the p-**regular factor** of g and b is the p-**singular factor** of g. If $b = 1$, i.e., if the order of g is prime to p, then g is a p-**regular element** of G. If $a = 1$, i.e., if the order of g is a power of p, then g is a p-**singular element** of G.

Lemma 1.8. *Let $L \subset \mathbf{F}(G)$ be the $\mathbf{F}$-vector space spanned by $\{g_1 g_2 - g_2 g_1 \mid g_1, g_2 \in G\}$. Let $L_0 = \{\sum_{g \in G} a_g g \mid \sum_{g \in C_i} a_g = 0 \text{ for each } i\}$, where $\{C_i\}$ are the conjugacy classes of G. Then $L = L_0$.*

Proof. Since $g_1 g_2$ and $g_2 g_1$ are conjugate in G, for any g_1, g_2, we have that $g_1 g_2 - g_2 g_1 \in L_0$ and hence $L \subseteq L_0$.

On the other hand let $\sum_{g \in G} a_g g = \sum_i \sum_{g \in C_i} a_g g$ be in L_0. Fix $c_i \in C_i$. Then, by the definition of L_0, $\sum_{g \in C_i} a_g g = \sum_{g \in C_i} a_g(g - c_i)$ for each i. But if $g \in C_i$, $g = h c_i h^{-1}$ for some $h \in G$, and then $g - c_i = (h c_i)h^{-1} - h^{-1}(h c_i) \in L$ and hence $L_0 \subseteq L$. $\square$

Theorem 1.9 (Brauer). *Let $\mathbf{F}$ be a splitting field for G. Then the number of distinct irreducible $\mathbf{F}$-representations of G is equal to t', the number of p-regular conjugacy classes of G, (i.e., the number of conjugacy glasses of p-regular elements of G).*

Proof. Let $C_1,\dots,C_t$ be the conjugacy classes of G, and order them so that $C_1,\dots,C_{t'}$ are the conjugacy classes of p-regular elements of G. Fix $c_i \in C_i$, $i = 1,\dots,t$. Let $C_1 = \{1\}$, so $c_1 = 1$.

We shall show $\mathcal{C}' = \{c_i\}$, $i = 1,\dots,t'$, is a basis for R/M, where M is as in Lemma 6.4.8.

First we show $\mathcal{C}'$ spans. Let $\mathcal{C} - \{c_i\}$, $i = 1,\dots,t$. As we have observed in the proof of Lemma 1.8, if g is conjugate to c_i, then $g \equiv c_i \pmod{L}$. Since $\{g \mid g \in G\}$ spans R, this shows $\mathcal{C}$ spans R/L, and hence R/M.

Now suppose c_i is not p-regular. Write $c_i = ab$ as in Lemma 1.6, so $b^{p^j} = 1$ for some j. Then $(c_i - a)^{p^j} = 0$, so $c_i - a \in J(R) \subseteq M$, so $c_i \equiv a \pmod{M}$, and $\mathcal{C}'$ spans, as claimed.

Next we show $\mathcal{C}'$ is linearly independent. Let $\sum_{i=1}^{t'} a_i c_i \in M$. Then $\left(\sum_{i=1}^{t'} a_i c_i\right)^{p^j} \in L$ for some j. Since each c_i has order relatively prime to p, there is a $k \geq j$ with $c_i^{p^k} = c_i$ for each i. Then $0 \equiv \left(\sum_{i=1}^{t'} a_i c_i\right)^{p^k} \equiv \sum_{i=1}^{t'} a_i^{p^k} c_i \pmod{L}$. But now, by Lemma 1.8, $a_i^{p^k} = 0$ for each i, so $a_i = 0$ for each i. $\qquad\square$

Theorem 1.10. *Let H be a subgroup of G.*

(1) *If W is a projective $\mathbf{F}(H)$-module, then $\mathrm{Ind}_H^G(W)$ is a projective $\mathbf{F}(G)$-module.*

(2) *If V is a projective $\mathbf{F}(G)$-module, then $\mathrm{Res}_H^G(V)$ is a projective $\mathbf{F}(H)$-module.*

Proof.

(1) Let $W \oplus W' \cong k\mathbf{F}(H)$. Then $\mathrm{Ind}_H^G(W) \oplus \mathrm{Ind}_H^G(W') \cong k\,\mathrm{Ind}_H^G(\mathbf{F}(H)) = k\mathbf{F}(G)$.

(2) Let $V \oplus V' \cong k\mathbf{F}(G)$. Then $\mathrm{Res}_H^G(V) \oplus \mathrm{Res}_H^G(V') \cong k\,\mathrm{Res}_H^G(\mathbf{F}(G)) = k[G : H]\mathbf{F}(H)$. $\qquad\square$

Lemma 1.11. *Let $\mathbf{F}$ be a splitting field for G.*

(1) *For each i, $x_i = \dim_{\mathbf{F}}(S_i)$ is divisible by p^r.*

(2) **(Dickson)** *For each i, the number of times y_i that T_i appears in a composition series for $\mathbf{F}(G)$ is divisible by p^r.*

(3) *For each i, $x_i = y_i$.*

Proof. (1) Since S_i is a projective $\mathbf{F}(G)$-module, $\mathrm{Res}_P^G(S_i)$ is a projective $\mathbf{F}(P)$-module, so by Proposition 1.1 and the Krull-Schmidt theorem (Theorem 6.1.19 (2)) it must be isomorphic to $k_i\mathbf{F}(P)$, of dimension $k_i p^r$, for some k_i.

(2) and (3) By Proposition 6.3.14, the number of times T_i appears in a composition series for $\mathbf{F}(G)$ is equal to $\dim_{\mathbf{F}} \mathrm{Hom}_{\mathbf{F}(G)}(S_i, \mathbf{F}(G))$. But, by Frobenius reciprocity,

$$
\begin{aligned}
\mathrm{Hom}_{\mathbf{F}(G)}(S_i, \mathbf{F}(G)) &= \mathrm{Hom}_{\mathbf{F}(G)}\big(S_i, \mathrm{Ind}_P^G(\mathbf{F}(P))\big) \\
&\cong \mathrm{Hom}_{\mathbf{F}(P)}\big(\mathrm{Res}_P^G(S_i), \mathbf{F}(P)\big) \\
&\cong \mathrm{Hom}_{\mathbf{F}(P)}\big(k_i\mathbf{F}(P), \mathbf{F}(P)\big) \\
&\cong k_i\,\mathrm{Hom}_{\mathbf{F}(P)}\big(\mathbf{F}(P), \mathbf{F}(P)\big) \\
&\cong k_i\mathbf{F}(P)
\end{aligned}
$$

with k_i as in the proof of part (1), and $\dim_{\mathbf{F}} k_i\mathbf{F}(P) = k_i p^r$. $\qquad\square$

Proposition 1.12. *Let $\mathbf{F}$ be a splitting field for G. Let $G/[G:G]$ have order $n' = p^{r'}q'$ with $(p, q') = 1$. Then the number of distinct 1-dimensional $\mathbf{F}$-representations of G is equal to q'.*

Proof. Exactly the same as the proof of Proposition 3.2.5. $\qquad\square$

The technique of averaging over the group (Lemma 3.1.13) was crucial in proving Maschke's theorem (Theorem 3.1.14) about semisimplicity of $\mathbf{F}(G)$. Here $\mathbf{F}(G)$ is not semisimple, but we have a relative version of this technique.

Lemma 1.13. *Let H be a subgroup of G with $[G:H] = k$ relatively prime to p. Let $\{g_j\}$ be a set of right coset representatives of H. Let V_1 and V_2 be $\mathbf{F}$-representations of G defined by $\sigma_i : G \to \mathrm{Aut}(V_i)$ for $i = 1, 2$. Let $f \in \mathrm{Hom}_H(V_1, V_2)$ and set*

$$
\mathrm{Av}(f) = \frac{1}{k} \sum_{g_j} \sigma_2(g_j^{-1}) f \sigma_1(g_j).
$$

Then $\mathrm{Av}(f) \in \mathrm{Hom}_G(V_1, V_2)$. Furthermore, if $f \in \mathrm{Hom}_G(V_1, V_2)$, then

$$
\mathrm{Av}(f) = f.
$$

Proof. Virtually identical to the proof of Lemma 3.1.13. $\qquad\square$

Corollary 1.14. *Let H be a subgroup of G with $[G:H]$ prime to p. Let $0 \to V_1 \to V \to V_2 \to 0$ be an exact sequence of $\mathbf{F}$-representations of G, and suppose that $0 \to \mathrm{Res}_H^G(V_1) \to \mathrm{Res}_H^G(V) \to \mathrm{Res}_H^G(V_2) \to 0$ splits as a sequence of $\mathbf{F}$-representations of H. Then the original sequence splits.*

Proof. Virtually identical to the proof of Theorem 3.1.14. $\qquad\square$

Let us see some applications of this result. Let H be any subgroup of G and W any **F**-representation of H. Then, by Lemma 4.1.4 (2), W is a summand of $\operatorname{Res}_H^G \operatorname{Ind}_H^G(W)$. We consider the other direction.

Corollary 1.15. *Let H be a subgroup of G with $[G : H]$ prime to p, and let V be any **F**-representation of G. Then V is a summand of $\operatorname{Ind}_H^G \operatorname{Res}_H^G(V)$.*

Proof. Let $\{g_i\}$ be a set of left coset representatives of H, with $g_1 = 1$. Then $V' = \operatorname{Ind}_H^G \operatorname{Res}_H^G(V) = \{\sum_i g_i \otimes v_i \mid v_i \in V\}$. Define $\varphi : V' \to V$ by $\varphi(\sum g_i \otimes v_i) = \sum g_i v_i$. This is an epimorphism, so we have an exact sequence $0 \to \operatorname{Ker}(\varphi) \to V' \to V \to 0$. Let $\alpha : V \to V'$ by $\alpha(v) = 1 \otimes v$. Then α is a splitting of φ, and α is a map of **F**-representations of H, so φ has a splitting by Corollary 1.14. (Indeed, this splitting is $\operatorname{Av}(\alpha)$.) $\square$

Theorem 1.16 (Higman; Kasch, Kneser, and Kupich).

(1) *If P is not cyclic, G has indecomposable **F**-representations of arbitrarily large degree. In particular, in this case G has infinitely many distinct indecomposable **F**-representations.*

(2) *If P is cyclic, G has at most n distinct indecomposable **F**-representations, and each of these is of degree at most n.*

Proof. First suppose P is not cyclic. Then P has a quotient isomorphic to $A = \mathbb{Z}/p\mathbb{Z} \times \mathbb{Z}/p\mathbb{Z}$. (Proof: By induction on the order of P. If P is abelian, this is obvious, so suppose not. Let C be the center of P. Then P/C cannot be cyclic, for that would make P abelian, a contradiction, so by induction P/C has a quotient isomorphic to A, i.e., $(P/C)/\overline{H}$ is isomorphic to A for some subgroup $\overline{H}$ of P/C, and so P/H is isomorphic to A where H is the inverse image of $\overline{H}$ in P.)

Let $\pi : P \to A$ be the quotient map. Let $\{M_m\}$ be the indecomposable representations of A constructed in Example 5.1.4. Let $W_m = \pi^*(M_m)$ and $V_m = \operatorname{Ind}_P^G(W_m)$. Write V_m as a direct sum of indecomposable **F**-representations of G, $V = V_m^1 \oplus \cdots \oplus V_m^k$. By Lemma 4.1.4 (2), W_m is a direct summand of $\operatorname{Res}_P^G(V_m)$. Since W_m is an indecomposable **F**-representation of P, W_m is a summand of $\operatorname{Res}_P^G(V_m^i)$ for some i, and so $\deg(V_m^i) \geq \deg(W_m) = \deg(M_m) = 2m + 1$, as required.

Next suppose P is cyclic. Let V be an indecomposable **F**-representation of G. Then V is a summand of $\operatorname{Ind}_P^G \operatorname{Res}_P^G(V)$ by Corollary 1.15, so is a summand of $\operatorname{Ind}_P^G(W)$ for some indecomposable representation W of P. But there are only finitely many of these (by Example 5.1.1), showing that there are only finitely many distinct indecomposable **F**-representations of G in this case. Indeed, by Example 5.1.1, every indecomposable **F**-representation of P is of degree at most $|P|$, so every indecomposable representation of G is of degree at most $|G|$.

To obtain an upper bound on the number of distinct indecomposable $\mathbf{F}$-representations of G we must do more work. Again, by Example 5.1.1, we see that P has a unique (up to isomorphism) indecomposable $\mathbf{F}$-representation W_i of degree i, for each i with $1 \leq i \leq p^r$, and no others. For each such i, let $V_i = \operatorname{Ind}_P^G(W_i)$ and let

$$\mathcal{V}_i = \{\text{indecomposable } \mathbf{F}\text{-representations of } G \text{ of degree at least } i$$
$$\text{that are direct summands of } V_i\}.$$

Since $\deg(V_i) = [G : P]\deg(W_i) = [G : P]i$, we see that $|\mathcal{V}_i| \leq [G : P]$ and so $\left|\bigcup_{i=1}^{p^r} \mathcal{V}_i\right| \leq [G : P]p^r = n$.

Now let V be any indecomposable $\mathbf{F}$-representation of G. We show that $V \in \mathcal{V}_i$ for some i, and this completes the proof. To see this, note by Corollary 1.15 that V is a direct summand of $\operatorname{Ind}_P^G \operatorname{Res}_P^G(V)$, so is a direct summand of $\operatorname{Ind}_P^G(W)$ for some indecomposable $\mathbf{F}$-representation W of P that is a direct summand of $\operatorname{Res}_P^G(V)$, i.e., for some indecomposable $\mathbf{F}$-representation W of P with $\deg(W) \leq \deg(V)$, so if $i = \deg(W)$, $V \in \mathcal{V}_i$. $\qquad\square$

7.2. Characters and Brauer Characters

In this section we investigate characters of modular representations, and define related objects, known as Brauer (or modular) characters, which yield additional information.

Theorem 2.1. *Let $\mathbf{F}$ be algebraically closed. Then the characters of the distinct irreducible representations of G are linearly independent as functions on G.*

Proof. The character of a representation is the sum of the diagonal entries in any matrix form of it, so this is immediate from the Frobenius-Schur theorem (Theorem 2.5.6), which we have observed also holds in the modular case (Theorem 6.4.10). $\qquad\square$

Lemma 2.2. *Let $\mathbf{F}$ be algebraically closed. Let V be an $\mathbf{F}$-representation of G, given by $\sigma : G \to \operatorname{Aut}(V)$. Let g be an element of G and write $g = ab$ as in Lemma 1.6. Then the eigenvalues of $\sigma(g)$ and $\sigma(b)$ are the same with the same multiplicities.*

Proof. Chose a basis in which $\sigma(g)$ is in Jordan canonical form. Since a and b are powers of g, $\sigma(a)$ and $\sigma(b)$ are upper triangular, and since the order of a is a power of p, then diagonal entries of $\sigma(a)$ are all equal to 1. But $\sigma(g) = \sigma(a)\sigma(b)$, so the result follows. $\qquad\square$

Corollary 2.3. *Let* **F** *be algebraically closed. The characters of the distinct irreducible* **F**-*representations of G form a basis for the space of class functions on the p-regular elements of G.*

Proof. Recall from Theorem 1.9 that the number of distinct irreducible **F**-representations of G is equal to t', the number of p-regular conjugacy classes. Then the result is immediate from Theorem 2.1 and Lemma 2.2. □

Theorem 2.4 (Brauer-Nesbitt). *Let V_1 and V_2 be semisimple representations of G over an algebraically closed field* **F** *of arbitrary characteristic, given by $\sigma_i : G \to \mathrm{Aut}(V_i)$, $i = 1, 2$. Then V_1 and V_2 are isomorphic if and only if $\sigma_1(g)$ and $\sigma_2(g)$ have the same eigenvalues with the same multiplicities or, equivalently, if $\sigma_1(g)$ and $\sigma_2(g)$ have the same characteristic polynomials, for every $g \in G$.*

Proof. The only-if part is trivial. The if part in characteristic 0 follows immediately from the fact that in this case V_1 and V_2 have the same characters.

Thus it remains to prove the if part when $\mathrm{char}(\mathbf{F}) = p > 0$. Let $\{W_i\}$ be the set of irreducible **F**-representations of G.

Suppose the theorem is false, and let V_1 and V_2 be counterexamples with $\dim_{\mathbf{F}} V_1 = \dim_{\mathbf{F}} V_2$ minimal.

Write $V_1 \cong \bigoplus m_i^1 W_i$ and $V_2 \cong \bigoplus m_i^2 W_i$ for some integers $\{m_i^1\}$ and $\{m_i^2\}$. Note by minimality that $\{i \mid m_i^1 \neq 0\}$ and $\{i \mid m_i^2 \neq 0\}$ are disjoint. Now, as a consequence of the hypothesis of the theorem, the representations V_1 and V_2 have the same characters. But the characters of distinct irreducible representations are linearly independent, regardless of characteristic, by Theorem 2.1. Hence we must have each m_i^1 and each m_i^2 divisible by p.

Write $U_1 = \bigoplus (m_i^1/p) W_i$ and $U_2 = \bigoplus (m_i^2/p) W_i$. Then U_1 and U_2 are a pair of nonisomorphic semisimple representations of G that are also counterexamples to the theorem. But this contradicts the minimality of $\dim_{\mathbf{F}} V_1 = \dim_{\mathbf{F}} V_2$. □

Recall we have Speiser's construction (Section 3.8) associating an $\overline{\mathbf{F}}_p$-representation $\overline{\varphi} = \mathrm{Red}_p(\varphi)$ of G to a $\overline{\mathbf{Q}}$-representation φ of G. There was a choice involved in the construction of $\overline{\varphi}$, and indeed, the isomorphism class of $\overline{\varphi}$ may depend on this choice. However, the type of $\overline{\varphi}$ (as defined in Definition 6.1.16) does not.

Corollary 2.5. *Let φ be a $\overline{\mathbf{Q}}$-representation of G. Then $\mathcal{T}(\mathrm{Red}_p(\varphi))$ is well defined. In particular, if $\mathrm{Red}_p(\varphi)$ is irreducible, its isomorphism class is well defined.*

Proof. The characteristic polynomial of $\varphi(g)$ determines the characteristic polynomial of $\overline{\varphi}(g)$, so the result follows immediately from Theorem 2.4. □

We remind the reader of the following result.

Theorem 2.6. *Let φ be a $\overline{\mathbf{Q}}$-representation of G of degree d. If p is prime to n/d, then $\mathrm{Red}_p(\varphi)$ is irreducible.*

Proof. This is just Theorem 3.8.2. $\qquad\qquad\square$

Remark. Set $\overline{\varphi} = \mathrm{Red}_p(\varphi)$, and suppose that $\mathcal{T}(\overline{\varphi}) = \{\psi_i\}$. Let the representation space of $\overline{\varphi}$ be V and the representation space of each ψ_i be W_i. Let $\psi = \bigoplus \psi_i$ with representation space $W = \bigoplus W_i$. Then, in an appropriate basis, for every g in G,

$$\overline{\varphi}(g) = \begin{bmatrix} \psi_1(g) & * & * & \\ 0 & \psi_2(g) & * & \\ 0 & 0 & \psi_3(g) & \\ & & & \ddots \end{bmatrix},$$

where the entries marked $*$ are not specified further. On the other hand,

$$\psi(g) = \begin{bmatrix} \psi_1(g) & 0 & 0 & \\ 0 & \psi_2(g) & 0 & \\ 0 & 0 & \psi_3(g) & \\ & & & \ddots \end{bmatrix}.$$

Thus Corollary 2.5 says that if V is given a basis in which $\overline{\varphi}$ is block upper triangular, the diagonal blocks of $\overline{\varphi}$ are well defined (i.e., independent of the choice of a basis for φ).

In the modular case, we have t conjugacy classes and t' absolutely irreducible representations. Lemma 2.2 shows that we do not lose any information if we restrict the characters of these t' representations to the t' p-regular conjugacy classes. The characters, of course, are $\mathbf{F}$-valued.

It turns out, however, that it is more fruitful to look at certain complex-valued class functions defined on the p-regular elements, known as modular characters or Brauer characters.

We begin by choosing, once and for all, an isomorphism α from the group of q-th roots of 1 in $\mathbf{F}$ to the group of q-th roots of 1 in $\mathbf{C}$.

Definition 2.7. Let V be an $\mathbf{F}$-representation of G given by $\sigma : G \to \mathrm{Aut}(V)$. The **Brauer character** β_V is the complex-valued class function on p-regular elements of G defined as follows. If $\{x_i\}$ are the eigenvalues of $\sigma(g)$ (each of which is a q-th root of 1 in $\mathbf{F}$) with multiplicities, then

$$\beta_V(g) = \sum \alpha(x_i).$$

Remark. Of course $\chi_V(g) = \sum x_i$. But we can see immediately that β_V gives more information than χ_V. For example, if $V = p\tau$, then $\beta_V(1) = p \in \mathbf{C}$, while $\chi_V(1) = 0 \in \mathbf{F}$.

Remark 2.8. For the proper choice of α, if $\overline{\varphi} = \mathrm{Red}_p(\varphi)$, then $\beta_{\overline{\varphi}} = \chi_\varphi$ as functions on the p-regular elements of G.

Theorem 2.9 (Brauer-Nesbitt). *Let V_1 and V_2 be $\mathbf{F}$-representations of G, where $\mathbf{F}$ is a splitting field for G. Then $\mathcal{T}(V_1) = \mathcal{T}(V_2)$, i.e., V_1 and V_2 have the same irreducible components with the same multiplicities, if and only if $\beta_{V_1}(g) = \beta_{V_2}(g)$ for every p-regular element g of G.*

Proof. The only-if part is trivial.

As for the if part, observe that for any $\mathbf{F}$-representation V of G, if $\beta_V(g) = \sum \alpha(x_i)$ as in Definition 2.7, then $\beta_V(g^r) = \sum \alpha(x_i^r) = \sum \alpha(x_i)^r$ for every r, so, by considering the elementary symmetric polynomials in $\{\alpha(x_i)\}$, we see that $\{\beta_V(g^r) \mid r = 1, 2, \ldots\}$ determines $\{\alpha(x_i)\}$, and hence $\{x_i\}$, and then the theorem follows from Lemma 2.2 and Theorem 2.4. $\square$

7.3. Examples

In this section we investigate the modular representations of a number of groups.

Example 3.1. Let $G = D_{2p}$, the dihedral group of order $2p$, p an odd prime, $G = \langle x, y \mid x^p = 1, y^2 = 1, xy = yx^{-1} \rangle$.

G has conjugacy classes $\{1\}$, $\{x^i, x^{p-i}\}$, $i = 1, \ldots, (p-1)/2$, $\{x^i y\}$.

Mod 2: G has $1 + (p-1)/2$ 2-regular conjugacy classes. Its irreducible representations are τ and $\mathrm{Red}_2(\varphi_i)$, $i = 1, \ldots, (p-1)/2$, in the notation of Example 3.1.4 (7). It is easy to check directly that $\mathrm{Red}_2(\varphi_i)$ is irreducible, though that also follows from Theorem 2.6, and that they are distinct, though that also follows from Remark 2.8 and Theorem 2.9.

Mod p: G has two p-regular conjugacy classes. Its irreducible representations are τ and ψ_-. Note that P is normal in G (P being the subgroup generated by x) and $G/P \cong \mathbb{Z}/2\mathbb{Z}$ so this verifies Corollary 1.3. Also, $\mathcal{T}(\mathrm{Red}_p(\varphi_i)) = \{\tau, \psi_-\}$. This is easy to check directly, though it also follows from computing Brauer characters.

Example 3.2. Let $G = D_{4p}$, the dihedral group of order $4p$, p an odd prime, $G = \langle xy \mid x^{2p} = 1, y^2 = 1, xy = yx^{-1} \rangle$.

G has conjugacy classes $\{1\}$, $\{x^i, x^{2p-i}\}$, $i-1,\ldots,p-1$, $\{x^p\}$, $\{x^{\text{even}}y\}$, $\{x^{\text{odd}}y\}$.

Mod 2: G has $1 + (p-1)$ 2-regular conjugacy classes. Its irreducible representations are τ and $\text{Res}_2(\varphi_i)$, $i = 1,\ldots,p-1$, as in Example 3.1.

Mod p: G has four 2-regular conjugacy classes. Its irreducible representations are $\tau = \psi_{++}, \psi_{+-}, \psi_{-+}, \psi_{--}$. Note also that P is normal in G (P being the subgroup generated by x^2) and $G/P \cong \mathbb{Z}/2\mathbb{Z} \oplus \mathbb{Z}/2\mathbb{Z}$, so this verifies Corollary 1.3. Note also that $\mathcal{T}(\text{Red}_p(\varphi_i)) = \{\tau, \psi_{+-}\}$.

Example 3.3. Let $G = S_3$, the symmetric group on three symbols. G has conjugacy classes $\{1\}$, $\{2\text{-cycles}\}$, $\{3\text{-cycles}\}$ and irreducible representations in characteristic 0 given by τ, *sign* (the 1-dimensional representation that is multiplication by the sign of a permutation), and φ, a 2-dimensional representation given by permutation of coordinates of elements of $V = \{[v_1, v_2, v_3] \mid v_1 + v_2 + v_3 = 0\}$. (Recall that $[v_1, v_2, v_3]$ denotes the space of column vectors.) We consider this last representation further.

Note that V has basis $\{b_1, b_2\}$, where $b_1 = [1, -1, 0]$ and $b_2 = [0, 1, -1]$. Also, if we set $b_3 = [-1, 0, 1]$, then $b_3 = -(b_1 + b_2)$. The action of generators of G is given by the following.

$$(123) : b_1 \mapsto b_2 \qquad\qquad (12) : b_1 \mapsto -b_1$$
$$b_2 \mapsto -(b_1 + b_2) \qquad\qquad b_2 \mapsto (b_1 + b_2).$$

In this basis these elements have the following matrices.

$$(123) : \begin{bmatrix} 0 & -1 \\ 1 & -1 \end{bmatrix} \qquad (12) : \begin{bmatrix} -1 & 1 \\ 0 & 1 \end{bmatrix}.$$

Mod 2: G has two 2-regular conjugacy classes. One of its irreducible representations is $\tau = \text{Red}_2(sign)$. The other, by Theorem 2.6, is $\text{Red}_2(\varphi)$. In the basis $\{b_1, b_2\}$, these elements have the following matrices.

$$(123) : \begin{bmatrix} 0 & 1 \\ 1 & 1 \end{bmatrix} \qquad (12) : \begin{bmatrix} 1 & 1 \\ 0 & 1 \end{bmatrix}.$$

It is easy to see from this directly that $\text{Red}_2(\varphi)$ is indeed irreducible.

Mod 3: G has two 3-regular conjugacy classes. Its irreducible representations are τ and $\text{Red}_3(sign)$. If $\overline{V}$ is the mod 3 reduction of V, note that $\overline{V}$ has a subspace $\overline{V}_0 = \{[v_1, v_2, v_3] \mid v_1 = v_2 = v_3\}$ on which G acts trivially. If $\{\overline{b}_1, \overline{b}_2\}$ is the mod 3 reduction of $\{b_1, b_2\}$, then $\overline{V}_0$ has basis $[1, 1, 1] = \overline{b}_1 - \overline{b}_2$. Note that (computing mod 3):

$$(123) : \overline{b}_1 \mapsto \overline{b}_1 - (\overline{b}_1 - \overline{b}_2) \qquad (12) : \overline{b}_1 \mapsto -\overline{b}_1$$
$$\overline{b}_2 \mapsto \overline{b}_2 - (\overline{b}_1 - \overline{b}_2) \qquad \overline{b}_2 \mapsto -\overline{b}_2 + (\overline{b}_1 - \overline{b}_2)$$

and $\overline{V}/\overline{V}_0$ has basis $\{\overline{b}_1\}$ (or $\{\overline{b}_2\}$) so we see this action is *sign*, and hence $\mathcal{T}(\mathrm{Red}_3(\varphi)) = \{\tau, sign\}$.

Actually, we can see this easily using Brauer characters. The Brauer character table for S_3 mod 3 is as follows.

	1	(12)
τ	1	1
$sign$	1	-1

The Brauer character β of $\mathrm{Red}_3(\varphi)$ is given by $\beta(1) = 2$, $\beta((12)) = 0$ (as $\chi_\varphi(1) = 2$ and $\chi_\varphi((12)) = 0$), and this is the sum of the Brauer characters of τ and of *sign*.

Example 3.4. Let $G = A_4$, the alternating group on four symbols. G has conjugacy classes $\{1\}$, {elements of order 2}, $\frac{1}{2}${elements of order 3}, $\frac{1}{2}${elements of order 3} (i.e., the elements of order 3 break up into two conjugacy classes). G is a semidirect product $1 \to V \to A_4 \to S \to 1$ with V the (normal 2-Sylow) subgroup isomorphic to $\mathbb{Z}/2\mathbb{Z} \oplus \mathbb{Z}/2\mathbb{Z}$ and S isomorphic to $\mathbb{Z}/3\mathbb{Z}$. G has four complex irreducible representations of degrees 1, 1, 1, 3.

Mod 2: G has three 2-regular conjugacy classes. Since V is normal, by Corollary 1.3 the irreducible representations of G are the pullbacks of the irreducible representations of G/V. Hence G has three 1-dimensional (irreducible) representations, τ, $\overline{\alpha}'_1$, $\overline{\alpha}''_1$, given by letting V act trivially and by letting a fixed generator of S act by multiplication by any of the three cube roots of 1 in $\overline{\mathbf{F}}_2$.

The Brauer character table of G mod 2 is as follows.

	(1)	(123)	(132)
τ	1	1	1
$\overline{\alpha}'_1$	1	ζ	ζ^2
$\overline{\alpha}''_1$	1	ζ^2	ζ

If α is the irreducible complex representation of G of degree 3, with $\chi_\alpha(1) = 3$, $\chi_\alpha((12)) = -1$, $\chi_\alpha((123)) = 0$, $\chi_\alpha((132)) = 0$, then its Brauer character β is given by $\beta(1) = 3$, $\beta((123)) = 0$, $\beta((132)) = 0$, so we see $\mathcal{T}(\mathrm{Red}_2(\alpha)) = \{\tau, \overline{\alpha}'_1, \overline{\alpha}''_1\}$.

Mod 3: G has two 3-regular conjugacy classes, and hence two irreducible representations. One of these is τ. (Note that all three 1-dimensional complex representations reduce to τ.) The other, by Theorem 2.6, is $\mathrm{Red}_3(\alpha)$, where α is the irreducible complex representation of degree 3. (Note that this representation is the natural action of A_4 by permutation of coordinates on $\{[v_1, v_2, v_3, v_4] \mid v_1 + v_2 + v_3 + v_4 = 0\}$.)

Example 3.5. Let $G = S_4$, the symmetric group on four symbols. G has conjugacy classes $\{1\}$, $\{2\text{-cycle}\}$, $\{(2\text{-cycle})(2\text{-cycle})\}$, $\{3\text{-cycle}\}$, $\{4\text{-cycle}\}$. G is a semidirect product $1 \to V \to S_4 \to T \to 1$ with V isomorphic to $\mathbb{Z}/2\mathbb{Z} \oplus \mathbb{Z}/2\mathbb{Z}$ and T isomorphic to S_3. Let $\pi : G \to T$ be the quotient map. G has five irreducible complex representations of degrees 1, 1, 2, 3, 3.

Mod 2: G has two 2-regular conjugacy classes. Since V is normal, by Lemma 1.2 it acts trivially on every irreducible representation, so these factor through T. Thus, by Example 3.3, G has two irreducible representations, τ of degree 1, and $\mathrm{Red}_2(\pi^*(\varphi))$ of degree 2.

Clearly the two 1-dimensional complex representations τ and *sign* reduce to τ. The two irreducible 3-dimensional complex representations are α_3 and *sign* $\otimes \alpha_3$, so they clearly have the same reduction. Now α_3 is the representation of S_4 on $V = \{[v_1, v_2, v_3, v_4] \mid v_1 + \cdots + v_4 = 0\}$ by permuting coordinates. If $\overline{V}$ is the mod 2 reduction of V, then $\overline{V}$ has basis $\{\overline{b}_1, \overline{b}_2, \overline{b}_3\} = \{[1,1,0,0], [0,1,1,0], [0,0,1,1]\}$. Note that $\overline{V}$ contains the subspace $\overline{V}_0 = \{[v,v,v,v]\}$ on which G acts trivially, and $\overline{V}_0$ has basis $\{\overline{b}_1 + \overline{b}_3\} = \{[1,1,1,1]\}$, and $\overline{V}/\overline{V}_0$ has basis $\{\overline{b}_1, \overline{b}_2\}$. The action of generators of G on $\overline{V}/\overline{V}_0$ is

$$(123) : \overline{b}_1 \mapsto \overline{b}_2 \qquad (12) : \overline{b}_1 \mapsto \overline{b}_1 \qquad (12)(34) : \overline{b}_1 \mapsto \overline{b}_1$$
$$\overline{b}_2 \mapsto \overline{b}_1 + \overline{b}_2 \qquad \overline{b}_2 \mapsto \overline{b}_1 + \overline{b}_2 \qquad \overline{b}_2 \mapsto \overline{b}_2$$

so the action factors through T, and then these elements have the following matrices.

$$(123) : \begin{bmatrix} 0 & 1 \\ 1 & 1 \end{bmatrix} \qquad (12) : \begin{bmatrix} 1 & 1 \\ 0 & 1 \end{bmatrix}.$$

By comparison with Example 3.3 we see this representation is $\mathrm{Red}_2(\pi^*(\varphi))$. Hence $\mathcal{T}(\mathrm{Red}_2(\alpha_3)) = \{\tau, \mathrm{Red}_2(\pi^*(\varphi))\}$. (Again, we could compute this via Brauer characters.)

Mod 3: G has four 3-regular conjugacy classes. G has two 1-dimensional irreducible representations τ and $\mathrm{Red}_3(sign)$. By Theorem 2.6, $\mathrm{Red}_3(\alpha_3)$ and $\mathrm{Red}_3(\alpha_3')$ are each irreducible, and they are distinct as in the first of these the element (12) of G has character value 1 while in the second it has character value -1. As in Example 3.3, we see that the reduction of the 2-dimensional irreducible complex representation has type $\mathcal{T}(\mathrm{Red}_3(\pi^*(\varphi))) = \{\tau, sign\}$.

Example 3.6. Let $G = A_5$, the alternating group on five symbols. G has conjugacy classes $\{1\}$, $\{(2\text{-cycle})(2\text{-cycle})\}$, $\{3\text{-cycles}\}$, $\frac{1}{2}\{5\text{-cycles}\}$, $\frac{1}{2}\{5\text{-cycles}\}$, (i.e., the 5-cycles break up into two conjugacy classes) and has irreducible complex representations of degrees 1, 3, 3, 4, 5. For convenience, we reproduce its complex character table from Example 3.5.26 (see also Example 4.1.26). We let $\zeta = \exp(2\pi i/5)$ and let $c_i \in C_i$.

	C_1	C_2	C_3	C_4	C_5
$\alpha_1 = \tau$	1	1	1	1	1
α_3	3	-1	0	$1 + \zeta + \zeta^4$	$1 + \zeta^2 + \zeta^3$
α_3'	3	-1	0	$1 + \zeta^2 + \zeta^3$	$1 + \zeta + \zeta^4$
α_4	4	0	1	-1	-1
α_5	5	1	-1	0	0

Mod 2: G has four 2-regular conjugacy classes. G has irreducible representations τ and $\mathrm{Red}_2(\alpha_4)$ by Theorem 2.6, so we need to find two others. Let $\beta_1, \beta_3, \beta_3', \beta_4, \beta_5$ be the Brauer characters of the reductions of these five irreducible complex representations, and note that they are given by the above table with the second column deleted.

Note that α_3 and α_3' are conjugate representations, so their reductions are either both irreducible or both reducible. Note these reductions are not isomorphic as their Brauer characters differ. (Indeed, their characters also differ, as if $\overline{\zeta}$ is a primitive fifth root of 1 in $\overline{\mathbf{F}}_2$, $1 + \overline{\zeta} + \overline{\zeta}^4 \neq 1 + \overline{\zeta}^2 + \overline{\zeta}^3$ in $\overline{\mathbf{F}}_2$.) Observe that as $[G : G] = G$, G has only one 1-dimensional representation and that is τ. Now $\mathcal{T}(\mathrm{Red}_2(\alpha_3)) \neq \{\tau, \tau, \tau\}$ as their character (or Brauer character) values on C_3 differ, and similarly for $\mathcal{T}(\mathrm{Red}_2(\alpha_3'))$. Note also that $\beta_3 + \beta_3' = \beta_1 + \beta_5$. Thus, by Theorem 2.9, $\mathcal{T}(\mathrm{Red}_2(\alpha_3)) \cup \mathcal{T}(\mathrm{Red}_2(\alpha_3')) = \mathcal{T}(\mathrm{Red}_2(\alpha_3 \oplus \alpha_3')) = \mathcal{T}(\mathrm{Red}_2(\alpha_1 \oplus \alpha_5)) = \{\tau\} \cup \mathcal{T}(\mathrm{Red}_2(\alpha_5))$. This is impossible if $\mathrm{Red}_2(\alpha_3)$ and $\mathrm{Red}_2(\alpha_3')$ are irreducible. Hence they must be reducible, and we must have that

$$\mathcal{T}(\mathrm{Red}_2(\alpha_3)) = \{\tau, \overline{\alpha}_2\},$$
$$\mathcal{T}(\mathrm{Red}_2(\alpha_3')) = \{\tau, \overline{\alpha}_2'\},$$
$$\mathcal{T}(\mathrm{Red}_2(\alpha_5)) = \{\tau, \overline{\alpha}_2, \overline{\alpha}_2'\}$$

for some irreducible 2-dimensional representations $\overline{\alpha}_2$ and $\overline{\alpha}_2'$ of G (that are *not* reductions of complex representations).

The Brauer character table of A_5 mod 2 is then as follows.

	C_1	C_3	C_4	C_5
τ	1	1	1	1
$\overline{\alpha}_2$	2	-1	$\zeta + \zeta^4$	$\zeta^2 + \zeta^3$
$\overline{\alpha}_2'$	2	-1	$\zeta^2 + \zeta^3$	$\zeta + \zeta^4$
$\overline{\alpha}_4$	4	1	-1	-1

Mod 3: G has four 3-regular conjugacy classes. G has irreducible representations τ, $\mathrm{Red}_3(\alpha_3)$ and $\mathrm{Red}_3(\alpha_3')$, by Theorem 2.6, and these are again distinct as their characters (or Brauer characters) differ. Now α_4 is the action of A_5 by permuting coordinates in $\{[v_1, \ldots, v_5] \mid v_1 + \cdots + v_5 = 0\}$, and it is easy to check directly that $\mathrm{Red}_3(\alpha_4)$ is irreducible. Also, we have the equation in Brauer characters $\beta_5 = \beta_1 + \beta_4$, so $\mathcal{T}(\mathrm{Red}_3(\alpha_5)) = \{\tau, \mathrm{Red}_3(\alpha_4)\}$.

The Brauer character table of A_5 mod 3 is obtained from the ordinary character table by deleting the fifth row and third column.

Mod 5: G has three 5-regular conjugacy classes. G has irreducible representations τ and $\mathrm{Red}_5(\alpha_5)$, by Theorem 2.6. We claim $\mathrm{Red}_5(\alpha_3)$ is irreducible. Note that τ is the only 1-dimensional representation of G. Certainly $\mathcal{T}(\mathrm{Red}_5(\alpha_3)) \neq \{\tau, \tau, \tau\}$ as their character (or Brauer character) values on c_2 differ. Thus, if $\mathrm{Red}_5(\alpha_3)$ were reducible, the only possibility would be $\mathcal{T}(\mathrm{Red}_5(\alpha_3)) = \{\tau, \overline{\alpha}_2\}$ for some irreducible 2-dimensional representation $\overline{\alpha}_2$ and G. But the Brauer character of $\mathrm{Red}_5(\alpha_3)$ on c_2 is equal to -1, and that of τ on c_2 is equal to 1, so that would force the Brauer character of $\overline{\alpha}_2$ on c_2 to be -2. Now c_2 has order 2, so the eigenvalues of $\overline{\alpha}_2(c_2)$ must be ± 1. This means that $\overline{\alpha}_2(c_2)$ would have two eigenvalues of -1. But this element is diagonalizable (as its order, 2, is prime to the characteristic, 5), so that forces $\overline{\alpha}_2(c_2) = \begin{bmatrix} -1 & 0 \\ 0 & -1 \end{bmatrix}$, a central element in $\mathrm{Mat}_2(\overline{\mathbf{F}}_2)$. But the center of A_5 is trivial, so that means $\overline{\alpha}_2$ would have a nontrivial kernel. But A_5 is simple, so that means $\overline{\alpha}_2$ would be trivial, a contradiction (as $\overline{\alpha}_2(c_2) \neq \begin{bmatrix} 1 & 0 \\ 0 & 1 \end{bmatrix}$).

Thus the irreducible representations of G are τ, $\mathrm{Red}_5(\alpha_3)$, and $\mathrm{Red}_5(\alpha_5)$. Note that $\beta_3 = \beta_3'$ so $\mathrm{Red}_5(\alpha_3') = \mathrm{Red}_5(\alpha_3)$. Also, $\beta_4 = \beta_1 + \beta_3$ so $\mathcal{T}(\mathrm{Red}_5(\alpha_4)) = \{\tau, \mathrm{Red}_5(\alpha_3)\}$.

The Brauer character table of A_5 mod 5 is given by the ordinary character table with the third and fourth rows and the fourth and fifth columns deleted.

Some Useful Results

In this appendix we collect several results we have used earlier, but which are not results in representation theory.

Theorem 1.1 (Wedderburn). *Every finite division ring D is a field.*

Proof (Witt). Let the center of D be the field $\mathbf{F}_q$ of q elements, and let D have dimension k as an $\mathbf{F}_q$-vector space, so D has q^k elements. Consider the multiplicative group D^* of D, of order $q^k - 1$. For each $a \in D^*$, its centralizer $C(a) = \{b \in D^* \mid ab = ba\}$ is a subgroup of D^*, and also $C(a) \cup \{0\}$ is a sub-$\mathbf{F}_q$-vector space of D. Hence $C(a)$ must have order $q^j - 1$ with $q^j - 1$ a divisor of $q^k - 1$, and hence j a factor of k. Also, the conjugacy class of a in D^* contains $(q^k - 1)/(q^j - 1)$ elements. Then, writing D^* as a union of conjugacy classes, and counting elements, we see that

$$q^k - 1 = q - 1 + \sum \frac{q^k - 1}{q^{j_i} - 1},$$

where the summand $q - 1$ comes from the nonzero elements of $\mathbf{F}_q^*$, each of which is in a conjugacy class by itself.

Now let $\Phi_m(x)$ denote the m-th cyclotomic polynomial. Note that $\Phi_k(x)$ divides not only $x^k - 1$, but also $(x^k - 1)/(x^j - 1)$, as polynomials in $\mathbb{Z}[x]$, for any proper divisor j of k (as $\Phi_k(x)$ and $x^j - 1$ are relatively prime for any such j).

Thus $\Phi_k(q)$ divides every term in the above equation except possibly $q - 1$, so $\Phi_k(q)$ must divide $q - 1$ as well. But $\Phi_k(q) = \Pi(q - \zeta)$, where the product is taken over the primitive k-th roots of unity. However, if $k > 1$, $|q - \zeta| > |q - 1|$ for any primitive k-th root of unity ζ, a contradiction. Hence $k = 1$ and $D = \mathbf{F}_q$ is commutative. $\qquad\square$

Theorem 1.2.

(1) *Let* $\mathbf{E}$ *be any finite extension of* $\mathbf{Q}$, *and let* $\mathcal{O}$ *denote the ring of integers of* $\mathbf{E}$. *Then there is a finite extension* $\widetilde{\mathbf{E}}$ *of* $\mathbf{E}$, *with ring of integers* $\widetilde{\mathcal{O}}$, *such that for any ideal* $\mathcal{I}$ *of* $\mathcal{O}$, $\widetilde{\mathcal{I}} = \mathcal{I}\widetilde{\mathcal{O}}$ *is a principal ideal of* $\widetilde{\mathcal{O}}$.

(2) *Let* $\overline{\mathcal{I}}$ *be any finitely generated ideal in the ring of integers* $\overline{\mathcal{O}}$ *of* $\overline{\mathbf{Q}}$. *Then* $\overline{\mathcal{I}}$ *is principal.*

Proof.

(1) The class number, i.e., the order of the ideal class group of $\mathbf{E}$, is finite ([Sa, section 4.3]), so this group has finite exponent k. Then for any ideal $\mathcal{I}$ of $\mathcal{O}$, $\mathcal{I}^k$ is principal, i.e., $\mathcal{I}^k = y\mathcal{O}$ for some $y \in \mathcal{O}$.

Let $\widetilde{\mathbf{E}} = \mathbf{E}[x]/(x^k - y)$ and let $\widetilde{\mathcal{J}} = y\widetilde{\mathcal{O}}$. Then on the one hand

$$\widetilde{\mathcal{J}} = y\widetilde{\mathcal{O}} = x^k\widetilde{\mathcal{O}} = (x\widetilde{\mathcal{O}})^k,$$

and on the other hand

$$\widetilde{\mathcal{J}} = y\widetilde{\mathcal{O}} = \mathcal{I}^k\widetilde{\mathcal{O}} = (\mathcal{I}\widetilde{\mathcal{O}})^k.$$

But $\widetilde{\mathcal{O}}$ is a Dedekind ring and the set of nonzero ideals of a Dedekind ring has unique factorization ([Sa, Section 3.4]), so $\mathcal{I}\widetilde{\mathcal{O}} = x\widetilde{\mathcal{O}}$ and $\widetilde{\mathcal{I}} = \mathcal{I}\widetilde{\mathcal{O}}$ is principal.

(2) Let $\overline{\mathcal{I}}$ be generated by $\{y_1, \dots, y_\ell\}$. Then there is a finite extension $\mathbf{E}$ of $\mathbf{Q}$ containing $\{y_1, \dots, y_\ell\}$. Let $\mathcal{O}$ be the ring of integers in $\mathbf{E}$ and let $\mathcal{I}$ be the ideal in $\mathcal{O}$ generated by $\{y_1, \dots, y_\ell\}$. Then by (1) there is a finite extension $\widetilde{\mathbf{E}}$ of $\mathbf{E}$, with ring of integers $\widetilde{\mathcal{O}}$, such that $\widetilde{\mathcal{I}} = \mathcal{I}\widetilde{\mathcal{O}}$ is a principal ideal of $\widetilde{\mathcal{O}}$, i.e., $\widetilde{\mathcal{I}} = x\widetilde{\mathcal{O}}$ for some $x \in \widetilde{\mathcal{O}}$. But then $\overline{\mathcal{I}} = x\overline{\mathcal{O}}$ is a principal ideal of $\overline{\mathcal{O}}$. $\qquad\qquad\square$

Theorem 1.3.

(1) *Let* R *be a Prüfer ring, i.e., an integral domain with the property that every finitely generated ideal of* R *is projective. Let* M *be a finitely generated torsion-free* R-*module. Then* M *is projective.*

(2) *Let* R *be a Bézout domain, i.e., an integral domain with the property that every finitely generated ideal of* R *is free. Let* M *be a finitely generated torsion-free* R-*module. Then* M *is free.*

Proof. (1) Our proof here is an adaptation of the proof of [AW, Theorem 3.6.6]. We proceed by induction on the number of generators k of M.

If $k = 1$, then $M = \langle m \rangle$ for some $m \in M$ and then $M = \langle m \rangle \cong Rm/\operatorname{Ann}(m)$ and $\operatorname{Ann}(m) = \{0\}$ by hypothesis, so $M \cong Rm$ is free.

Now suppose that the theorem is true for all R-modules generated by fewer than k elements, and let M be generated by $\{x_1, \ldots, x_k\}$. Let

$$M_1 = \{x \in M \mid ax = bx_1 \text{ for some } a, b \in R, a \neq 0\}.$$

Then M/M_1 is torsion free and is generated by $\{x_2 + M_1, \ldots, x_k + M_1\}$. Then, by the inductive hypothesis, M/M_1 is projective, so the exact sequence

$$\{0\} \to M_1 \to M \to M/M_1 \to \{0\}$$

splits and $M \cong M/M_1 \oplus M_1$. To complete the proof, we must show that M_1 is projective. We have just seen that M_1 is a direct summand of M, so there is a projection from M onto M_1, and since M is finitely generated, we conclude that M_1 is finitely generated as well. Let $\{y_1, \ldots, y_\ell\}$ be a generating set for M_1 and suppose that $a_i y_i = b_i x_1$ with $a_i \neq 0$ for $1 \leq i \leq \ell$. Let $q_0 = a_1 \cdots a_\ell$.

Claim. *If $ax = bx_1$ with $a \neq 0$, then $a \mid bq_0$.*

To see this, write $x = \sum_{i=1}^{\ell}$ with $c_i \in R$. Then it is easy to check that

$$bq_0 x_1 = a\left(\sum_{i=1}^{\ell} c_i(q_0/a_i)b_i\right)x_1,$$

so

$$bq_0 = a\left(\sum_{i=1}^{\ell} c_i(q_0/a_i)b_i\right)$$

as M_1 is torsion-free.

Using this claim we can define a function $\varphi : M_1 \to R$ by $\varphi(x) = (bq_0)/a$ whenever $ax = bx_1$ for $a \neq 0$. It is straightforward to show that φ is a well-defined homomorphism (since R is an integral domain and M_1 is torsion-free), and furthermore that $\mathrm{Ker}(\varphi) = \{0\}$. Thus

$$M \cong \mathrm{Im}(\varphi).$$

But $\mathrm{Im}(\varphi)$ is a finitely generated R-submodule of R, i.e., a finitely generated ideal, so it is projective by hypothesis.

(2) The proof of this is identical to the proof of (1), replacing the word "projective" by "free" wherever it appears. $\qquad\square$

Remark. It is easy to check that a nonzero ideal in an integral domain R is free as an R-module if and only if it is a principal ideal. Thus a Bézout domain is equivalently defined to be an integral domain R with the property that every finitely generated R-ideal is principal. Further, a nonzero ideal in an integral domain R is projective as an R-module if and only if it is an invertible ideal ([AW, Theorem 3.5.11]). Thus a Prüfer ring is equivalently defined to be an integral domain R with the property that every finitely generated R-ideal is invertible.

Theorem 1.4 (Adjoint associativity of Hom and tensor product).
Let M be an (S, R)-bimodule, N a (T, S)-bimodule, and P a (T, U) bimodule. Then there is an (R, U)-bimodule isomorphism

$$\Phi : \mathrm{Hom}_S(M, \mathrm{Hom}_T(N, P)) \to \mathrm{Hom}_T(N \otimes_S M, P).$$

Proof. Define Φ by
$$\Phi(f)(n \otimes m) = (f(m))(n),$$
where $f \in \mathrm{Hom}_S(M, \mathrm{Hom}_T(N, P))$, $m \in M$, and $n \in N$. Define Ψ by

$$(\Psi(g)(m))(n) = g(n \otimes m),$$

where $g \in \mathrm{Hom}_T(N \otimes_S M, P)$, $m \in M$, and $n \in N$.

Now Φ and Ψ are inverses of each other, as follows. If $g = \Phi(f)$, then

$$\begin{aligned}
(\Psi(g)(m))(n) &= g(n \otimes m) \\
&= \Phi(f)(n \otimes m) \\
&= (f(m))(n),
\end{aligned}$$

so $\Psi(g) = f$, and if $f = \Psi(g)$, then

$$\begin{aligned}
\Phi(f)(n \otimes m) &= (f(m))(n) \\
&= (\Psi(g)(m))(n) \\
&= g(n \otimes m),
\end{aligned}$$

so $\Phi(f) = g$, as required. $\qquad\square$

Theorem 1.5. *Let $\mathbf{F}$ be a finite field of characteristic p, and $\mathbf{F}'$ an arbitrary field of characteristic p. Then*

$$\mathbf{F}' \otimes_{\mathbf{F}_p} \mathbf{F} \cong \bigoplus \mathbf{F}_i'',$$

where $\{\mathbf{F}_i''\}$ are fields of characteristic p.

Proof. As $\mathbf{F}$ is finite, it is a Galois extension of $\mathbf{F}_p$, hence the splitting field of an irreducible polynomial $f(x)$ with no repeated roots.

Now for any $\mathbf{F}_p$-algebra R, and any two-sided ideal I of R,

$$\mathbf{F}' \otimes_{\mathbf{F}_p} (R/I) \cong (\mathbf{F}' \otimes_{\mathbf{F}_p} R)/(\mathbf{F}' \otimes_{\mathbf{F}_p} I).$$

Applying that here, we have

$$\mathbf{F}' \otimes_{\mathbf{F}_p} \mathbf{F} \cong \mathbf{F}' \otimes_{\mathbf{F}_p} (\mathbf{F}_p[x]/f(x)\mathbf{F}_p[x]) \cong \mathbf{F}'[x]/f(x)\mathbf{F}'[x].$$

But in $\mathbf{F}'[x]$, $f(x)$ factors as $\prod g_i(x)$ where $\{g_i(x)\}$ are irreducible, without repeated roots, and nonassociated (as $f(x)$ has no repeated roots). We claim that

$$\mathbf{F}'[x]/f(x)\mathbf{F}'[x] \cong \bigoplus \mathbf{F}'[x]/g_i(x)\mathbf{F}'[x] = \bigoplus \mathbf{F}''_i.$$

To see this claim, let $\alpha : \mathbf{F}'[x] \rightarrow \bigoplus \mathbf{F}'[x]/g_i(x)\mathbf{F}'[x]$ be the direct sum of the quotient maps. By the Chinese remainder theorem for $\mathbf{F}'[x]$, given any polynomials $h_i(x) \in \mathbf{F}'[x]$, there is a polynomial $h(x)$ with $h(x) \equiv h_i(x) \pmod{g_i(x)}$ for each i, so α is onto, and also by the Chinese remainder theorem, $h(x)$ is unique mod $f(x)\mathbf{F}'[x]$, so $\mathrm{Ker}(\alpha) = f(x)\mathbf{F}'[x]$, completing the proof. $\qquad\square$

Bibliography

[AW] Adkins, W. A., and Weintraub, S. H., *Algebra: An Approach via Module Theory*. Graduate Texts in Mathematics **136**, Springer-Verlag, 1992 (corrected second printing, 1999).

[BT] Brauer, R., and Tate, J. On the characters of finite groups. *Ann. of Math.* **62** (1955) 1-7.

[CR] Curtis, C. W., and Reiner, I. *Representation Theory of Finite Groups and Associative Algebras*. Pure and Applied Mathematics, Vol. XI, Interscience Publishers, 1962.

[HK] Hoffman, K, and Kunze, R., *Linear Algebra*, second edition. Prentice Hall, 1971.

[La] Lam, T. Y. *A First Course in Noncommutative Rings*, second edition. Graduate Texts in Mathematics **131**, Springer-Verlag, 2001.

[Ma] Matsumura, H. *Commutative Ring Theory*, Cambridge Studies in Advanced Mathematics **8**, Cambridge University Press, 1986.

[Sa] Samuel, P. *Algebraic theory of numbers*, translated by A. Silberger. Houghton Mifflin, 1970.

[Se] Serre, J.-P. *Linear Representations of Finite Groups*, translated by L. L. Scott. Graduate Texts in Mathematics **42**, Springer-Verlag, 1977 (corrected third printing, 1986).

Index

Titles in This Series